WEEDY AQUATIC PLANTS:

THEIR UTILITY, MENACE AND MANAGEMENT

Title cover plate: Close-up of an infestation of waterhyacinth severely attacked by its bioagent, *Neochetina eichhorniae* Warner

Inside cover figure: ***Ipomoea carnea*** **Jacq.**

"The first concern of the aquatic weed scientists is that the public, alongwith many political leaders, do not show any great concern over the crisis created by aquatic weeds, particularly the exotic species. For whatever reasons, they are in a state of denial. But what use can we make of a lake covered entirely with topped-out hydrilla or with mat of waterhyacinth. We can't fish it. Recreational boating or swimming is impossible. Even irrigation and flood control systems are adversely affected. In addition, the native aquatic biodiversity is being displaced by the exotic species and the water-side property values are being lowered." Think!

[From: Aquaphyte, USA]

DEDICATED TO MANAGERS OF WATERBODIES

AQUATIC PLANTS:
THEIR UTILITY, MENACE AND MANAGEMENT

—— A Reference Cum Text Book——

O. P. Gupta

Agrobios (India)

Published by:
AGROBIOS (INDIA)
Agro House, Behind Nasrani Cinema
Chopasani Road, Jodhpur 342 002
Phone: 91-0291-2642319, Fax: 2643993
E. mail: agrobios@sify.com

AGROBIOS (INIDA)

ISBN No. ...7-2
ISBN No. ...057-4
Price: Rs. ...5.00

9 788177 540574
Rs. 2500.00 / US$ .00
Published by:
AGROBIOS (INDIA)

Published by: Dr. Updesh Purohit for Agrobios (India), Jodhpur
Lasertypeset at: Yashee Computers and Printers, Jodhpur
Cover Design by: Reena
Printed at: Babloo Offset, Jodhpur

Preface

The objective of this book is to introduce to readers the (i) multifarious problems created by excessive growth of aquatic plants in our canals, lakes, ponds, drainage channels, lowland paddies, and other aquatic situations, leading to huge economic losses to the Nation year after year and (ii) various options available with us to curb these to within harmless levels. Thus far, these aquatic plant management options have been developed and used on field scale only in certain advanced countries of the world. We, in India, have yet to think about a constructive strategy to deal with this enormous problem by suitable Research and Development (R&D) procedures.

Presently, the research on aquatic weed management conducted in India has been largely based on laboratory studies, with little, if any, long term field-testing and development in the problematic waterbodies. In other words, the combined R&D efforts are completely lacking, except in 1-2 cases of biological control of specific aquatic weeds. This is a very unfortunate state of affairs. We require in the country 2-3 specialised centers for research and development each in (i) biological control, (ii) mechanical control, and (iii) mass utilization of aquatic weeds, supported by several field testing stations in different Agricultural Universities. The present book provides adequate subject matter to make a concrete beginning on these lines in the country.

Further, it seems no Agricultural University in the country offers even one course on the management of excessive aquatic plants. It is high time that a beginning is made in creating a course on this burning problem so that at least some students get oriented towards devoting to aquatic weed management in right earnestness. The present book provides the latest text available for formulating and conducting a post-graduate level course on aquatic weed management.

Acknowledgments: My thanks are due to Dr. R. K. Malik, Professor Agronomy, HAU, Hissar; Dr. J. Sharma, Professor Agronomy, H.P.K.V.V., Palampur; Dr. M. Arumugam, Assoc. Professor Agronomy, TNAU, Coimbatore; Dr. A. S. Rao, Assoc. Professor Agronomy, APAU, Hyderabad; Dr. A. N. Giri, Agronomist MAU, Parbhani; Dr. A. Basit & Dr. A. K. Gogoi, Scientists, AAU, Jorhat; and Dr. R.S. Singh, Agronomist, OUAT, Bhubaneshwar for providing information on aquatic weed flora and problems of respective states.

Some of the plates and figures used in this book were provided by the IFAS Center for Aquatic Plants, University of Florida, Gainsville, USA; Dr. A. Basit Professor, AAU, Jorhat; and Director, NRC (WS), Jabalpur, which are all sincerely acknowledged.

26.01.2001

O.P.G.

Preface

Contents

Chapter 1

INTRODUCTION

"Perhaps by now we have known a lot about water, but our knowledge about waterbodies is still in infancy; we could live better if we knew more about them"

With its multifarious uses, water is one of the most precious gifts of nature without which no life could survive on earth. When our population was limited, its bountiful supplies seemed endlessly renewable. We could then afford to foul one water source, abandon it, and move to another. This, however, is no longer possible since the exponential growth rates of human population, industries, and farms have already reduced the availability of water to below its minimum *per capita,* daily requirement of about 23 litres. Consequently, any further expansion of our activity will depend upon how well are we able to prevent the loss and contamination of the available water resources.

Biologically water constitutes a highly dynamic ecosystem with its dissolved nutrients and gases, chiefly oxygen. It is postulated that the plant life on our planet started in aquatic environment in the form of unicellular phytoplanktons. Later, during the process of evolution the multicellular plant structures developed. Many of these continued to grow in water, while others turned amphibious or terrestrial. Our interest in this chapter is limited to the plants growing within the freshwater waterbodies, with some concern also with plants growing on shores and ditchbanks, as well as in lowland areas.

1.1: THE NEED OF AQUATIC PLANTS

The presence of plants, including algae, in waterbodies is very important for the conservation of solar energy as chemical energy and enriching these with fresh oxygen during the process of photosynthesis. All this is essential to support in turn the varied aquatic fauna from unicellular zooplanktons to the large carp fish, prawns, and like which form valuable food of man (and certain animals). The zooplanktons and the phytoplanktons act as primary food for the growth of fish. Thus, the aquatic plants trigger the whole lot of interdependent life cycle, giving waterbody a living status. Without aquatic plants, the waterbodies shall be just "wet deserts".

Many aquatic plants have played fascinating roles in the life of man from the primitive times in one form or the other. For instance, the fruits of *Trapa natans var. bispinosa* (*singhara*, *paniphal*, waterchestnut) and of *Nelumbo nucifera* (*kamal-gatta*); corms, rhizomes, and tubers of *Colocasia esculenta* (taro), *Eleocharis dulcis* (Chinese waterchestnut), *Nelumbo nucifera* (*kamalkakri* or *bhen*), *Scirpus grossus* and *S. lacustris* (*keseru*); and the seeds of *Nelumbo nucifera* (*makhana*) and *Euryale ferox* Salisb (prickly *makhana,* Gorgannut) are very well known for their delicacies and economic values in India and elsewhere. Besides, several aquatic plants like *Ipomoea aquatica, Nasturtium officinale, Neptunia oleracea, Limnocharis flava, Ottelia alismoides*, and *Ceratopteris thalictriodes* form nutritious pot herbs (Gupta, 1987; Mazid, 1992; and Satpathy, 1964). The little plants of duckweeds form preferred food of ducks, geese, swan, and poultry (Traux *et al.*, 1972). Pigs too consume and grow on some aquatic plants, particularly the waterhyacinth.

The tiny plants of *Azolla* sp. carry on them a symbiont alga, *Anabaena azollae*, which fixes atmospheric nitrogen. Azolla is, therefore, often grown by the lowland rice growers as a biofertilizer plant for supplying nitrogen to their paddy crops (Sevicham and Madhusoodan, 1995; and Subba Rao, 1984). *Zizania aquatica* (wildrice) provides edible grains for humans in difficult times. Several water plants make water clear. Stoneworts like *Chara* spp. are particularly effective in this respect. (Van and Coops, 1999).

Not only the water plants but also the shore and ditchbank plants have attracted man. A wetland plant, *Phragmites communis* (common reed) has long been used by certain tribes for thatching, preparing articles of peasant crafts, and as a ditchbank binder. *Scirpus* and *Typha* spp. also provide thatching material. In Gujarat and Rajasthan, *Typha* spp. (Cattails) support cottage industries producing baskets, ropes, screens, winnowing trays, and mats. Certain *Carex, Pondus*, and *Juncus* spp. also provide material for making mats. *Typha* mats coated with plastic resins are as strong as fibreglass. In USA waterhyacinth has been used to make crafts and furniture under the waterhyacinth utilization project (WHUP).

Coming back to the true water plants, of recent, waterhyacinth (*Eichhornia crassipes*) has been found capable of demolishing heavy metal pollution caused by industrial releases carrying Cd, Pb, Cr, Ni and the like in waterbodies. Similarly, this floating plant, alongwith other plants like hydrilla, salvinia, duckweeds, parrotfeather, and elodea are known to scavenge upon large quantities of N, P, and K from waterbodies overenriched with these elements due to regular discharge of night sewage in them (Akcin *et al.*, 1994; Coquery and Welbourn, 1994; Moenandir and Murgito, 1994; Reddy, 1984; Roger and Davis, 1972; Steward, 1970; and Zaranyika, *et al.*, 1994).

Then there are aquatic plants capable of indicating the presence of metals in the sediments of waterbodies. These are known as bioindicators (Grasmuch *et al.*, 1995). Certain aquatic plants also prove useful in monitoring phytotoxic substances present in water.

Besides all the above materialistic gains from aquatic plants, one cannot forget their aesthetic values (Plates 1.1 and 1.2). There will be hardly any person who is not delighted by the gorgeous flowers of lotus and the fragrance of water-lilies. Even some non-flowering aquatic plants provide pleasing look to the surroundings. In Kashmir lakes floating gardens made out of aquatic plants for vegetable cultivation have long been delighting the tourists. In fact, scattered aquatic plants provide a pleasing variation in waterbodies. Who does not like to look at a large lake or pond filled with clean water, with some aquatic plants floating on its surface and some others growing here and there under it; a variety of fish swimming through these; ducks floating on the water and swans and cranes walking around its shores; and banks covered with dense growth of some low growing grasses? This is an ideal aquatic situation which is rare to find these days.

1.2: EXPLOSION OF AQUATIC PLANTS

In a balanced ecosystem aquatic plant population is kept in check through interaction of several physical, chemical, and biological factors operating both within and outside the waterbody. But when for some reason this balance is disturbed, rapid changes in the entire web of life are inevitable, including the rampant growth of aquatic plants, which does more harm to the waterbody than good. Our concern in rest of the book will be with various aspects of such excessive growth of aquatic plants called *aquatic weeds*.

*Three major reasons for the occurrence of aquatic weeds in waterbodies in the country and elsewhere are as follows:-

---

* ***Note:*** *Strictly speaking, aquatic weeds are plants which grow within any waterbody and infest it to the extent of hindering its intended uses. The overgrowths of plants occurring on the shores and banks of waterbodies and hindering either their convenient approaches or damaging these in any other way are called aquatic weed allies, although in terms of weed management these are also considered under the banner of aquatic weed problems.*

Plate 1.1: Man has spent huge sums of money in erecting palatial buildings near waterbodies. But with time several species of aquatic plants have filled these waterbodies and ruined such ideas. *Top,* A weedfree waterbody surrounding a beautifully planned hotel at Udaipur. (Photo-author). *Bottom,* A building complex developed near a weedfree waterbody in the past, presently faces a weed-filled waterbody. This should draw serious attention of all concerned.

Thus, broadly speaking, aquatic weeds are plants growing within and/or around waterbodies and impairing their efficient utilisation in one way or the other.

Plate 1.2: There is probably no better recreational substitute in summer to swimming and bathing in clean waterbodies. Aquatic plants should be prevented from overgrowing and taking this pleasure out of our waterbodies. *Top*, Children having recreation in a weedfree waterbody at Udaipur. *Bottom*, A waterbody at Kota filled with floating weeds near the shores and submersed weeds towards its middle. No meaningful recreational use of this waterbody can be anticipated. Note heaps of submersed weeds taken out from it and collected on the shore. These heaps contained largely eelgrass (*Vallisneria* sp.) (Photos-author).

1. Overenrichment of waterbodies with plant nutrients reaching there through run-off from agricultural lands, as well as through release of town sewage loads in them.

2. Heavy silting of waterbodies from eroded soils of catchment areas usually caused by felling of trees and non-adoption of land capability classification norms.

3. Introduction of extragenous aquatic plant species from distant continents.

The first two causes lead to *accelerated ageing* of waterbodies. The waterbodies then become increasingly shallow and polluted with essential plant nutrients. Both these conditions are congenial to the growth of masses of aquatic plants as weeds. When this phenomenon continues over years, the ponds and lakes may turn into marshes and, eventually, into dry places. The pollution of water with excessive plant nutrients is called **eutrophication**.

When the added organic matter like town sewage or worn out masses of aquatic weeds decompose in waterbodies, it greatly raises the BOD. (biological oxygen demand) of the waterbody besides inducing emission of foul smelling gases like CH_4, H_2S, and CO_2. Such *in situ* decomposition of organic matter is called **putrefaction**. Putrefaction leads to eutrophication when plant nutrients are released from organic wastes which, in turn, aggravates the growth of aquatic weeds. Unless checked, this vicious circle continues forcing the abandonment of the waterbody.

The *third* cause of our present level of aquatic weeds concerns with the adventive or evasive aquatic plant species which grow abundantly in the new environment for yet another reason *i.e.* the absence of their natural enemies, chiefly arthropods and plant pathogens which they escaped while they were transported deliberately or accidentally from their places of origin to new continents. Further, canals, large lakes, and natural ponds often force heavy seepage, making huge areas around them waterlogged, where aquatic weeds like cattails (*Typha* spp.) and reeds (*Phragmites* spp.) flourish.

1.3: MENACE OF EXCESSIVE AQUATIC PLANTS

1.3.1: Menace To Pisciculture

Among the edible animals, fish is most efficient user of water in terms of conversion of plant proteins into animal proteins. To produce 1 kg protein through other animals we need 12 times as much water as is needed by fish. Aquatic weeds are responsible for the loss of tonnes of such protein-efficient animals annually in varied ways. In India, as also in many other countries, the weedy floating mats of *E. crassipes*, *Pistia stratiotes*, and *Salvinia* spp. are most damaging to fisheries. These plants cover the water surface to cause darkness underneath, adversely affect the limnology of water, and prevent normal gaseous exchange between the water and atmosphere. In darkness, neither the phytoplanktons--the primary food organisms of fish--can grow normally nor the fish can locate their food. The water becomes cooler than usual and its dissolved oxygen content is disturbed under the cover of weeds. Also, the weeds may release in water their respiratory by-products of toxic nature.

The submersed aquatic weeds also damage fish and fish food. They reduce the development of phytoplanktons by competing for plant nutrients. Also, the weeds make access to the food difficult for the fish by obstructing their free movement. The predatory fish flourish under the shelter of weeds and compete with the cultivated species.

The excessive growth of planktonic blue-green algae, known as waterbloom (= water-flowers), is very damaging to fish, probably much more than the macrophytic, weedy vegetation. The waterblooms sometimes develop very suddenly in the nutrient-rich waterbodies on bright, sunny days. It then imparts a dark, soup-green colour to the water. During the day its heavy photosynthetic activity elevates the dissolved O_2 levels of water to as much 10 times its normal concentration of about 4 ppm. This causes distension of stomach in fish which puts them under severe distress. At night, or in the event of a sudden overcast during the day, a reverse phenomenon is noted. The masses of waterbloom algae then die abruptly and decompose. This consumes the dissolved O_2 of water rapidly, leaving this time the fish under severe strain of oxygen deficiency (Krishnamoorthi and Visveswarna, 1965). Also, the decomposing algae may release certain toxic metabolites. Death or injury to fish can be caused both due to oxygen depletion and ingestion of the toxins. The dead fish can be found floating on the surface of water at night after a heavy waterbloom. (Besides fish, other aquatic animals, wild life, and livestock can also be killed by consuming the bloom-infested waters).

The decomposition of worn-out roots and leaves of other noxious aquatic vegetation can also raise the BOD of a waterbody and create oxygen deficiency. However, this is a slow process and can not be fatal, although it may reduce fish growth.

Plate 1.3: A strong mat of aquatic plants like waterhyacinth (*Eichhornia crassipes*) or salvinia (*Salvinia* spp.) can stop even large navigation vessels. Here waterhyacinth mat is the menace. Courtesy: NAS, U.S.A.).

Plate 1.4: Several submersed aquatic plants wind around the propellers of motorboats and bring them to hault. This boat had to be towed to the shore for such reason. (Courtesy NAS, U.S.A.).

Plate 1.3

Plate 1.4

Scum algae can also sometimes cause considerable fish mortality by chocking their gills, besides inducing production of small fish.

Lastly, even if a crop of fish is made, aquatic weeds and algae may hinder their harvest. The fish nets and angling rods are often caught in aquatic weeds and are damaged.

1.3.2: Menace To Water Courses

Excessive, aquatic vegetation impedes the flow of water in all kinds of waterways, *e.g.* canals, flood control channels, and irrigation and drainage ditches. In India, weeds have already cut the water-flow in many canals by 20 to 95%. The direct effect of this is inadequate water supplies at the distal ends. Indirectly, the impediment of waterflow initiates the seepage process which turns vast areas of otherwise productive land around weedy canals waterlogged and, subsequently, saline. Silting of waterways induces floods in rivers. Too, the slow moving weedy water is subjected to accelerated evaporation in comparison to the fast moving one due to its longer exposure to the atmosphere.

The resistance to the flow of water in a stream offered by weeds or by its walls and bottom is measured as **retardant coefficient (n)**, using Manning's formulae. For the weedfree, good quality channels the `n' value is approximately 0.017-0.02, depending upon the nature of their substratum. In the presence of floating weeds, the 'n' value is raised to over twice that of the weedfree channels. The submersed weeds can increase the `n' value even 10 to 20-fold since they often rise in arcs to the surface of water and act as thousands of inverted combs. Mehta and Sharma (1976) found that in the distributaries of Chambal canals of Rajasthan, the submersed weeds (mainly *Potamogeton* and *Hydrilla* spp.) raised their Retardant Coefficients 3 to 5-folds. This amounted to a reduction in their water-flow by 60 to 80 per cent.

1.3.3: Menace To Water Passes And Intakes

The detached masses of aquatic weeds drift down the stream and clog culverts, bridge piers, pump intake points, hydroelectric plants, and dams. The tiny aquatic plants like *Lemna* and *Wolffia* spp., as well as scum alga present in agricultural watercourses block siphons, sprinklers, pumps, and sprayers.

1.3.4: Menace To Navigation

In rivers and large lakes, aquatic weeds keep back the pleasure of navigation and boating (Plates 1.3 and 1.4). They coil rou

0nd the propellers of motorboats and halt them. Mats of *E. crassipes* and *Salvinia* spp. can obstruct even large ships.

1.3.5: Menace To Potable Waterbodies

Under the cover of floating weeds like *Nymphaea, Nelumbo*, and *Eichhornia* spp., the drinking water becomes anaerobic and it remains inaccessible to man and his thirsty animals. Also, the decaying masses of aquatic weeds and algae pollute the water with odd metabolities and colouring matters and thus impart it disagreeable odour and taste. Algae also secrete some oily substances in water which clog the sand filters in the drinking water treatment plants.

Many aquatic plants elevate the evaporation loss of potable water, particularly during the summer months. Brezny *et al.* (1973) recorded water loss from tanks completely filled with, respectively, *E. crassipes, Typha* spp., and *Cyperus rotundus* as 30-50%, 60-70%, and 130-150% higher than evaporation from the weedfree tanks. *S. molesta* cover induced 1.5-3.7 times greater loss of water than evaporation from weedfree water surfaces.

Typha angustata grown in cement tanks was found to induce water loss that was 216% of the evaporation loss from similar, weedfree tanks (Mehta and Sharma, 1976).

1.3.6: Menace To Lowland Crops

Aquatic weeds invade and obstruct the growth of thousands of hectares of lowland paddy in West Bengal and other coastal parts of India. *E. crassipes, Salvinia* spp., and scum algae form prominent noxious aquatic vegetation destroying there the paddy fields. The scum algae (= slime) cause problems in rice fields by attaching themselves to the young crop seedlings and by forming mats that blanket the new, emerging rice shoots.

Cultivation of edible water-chestnut (*Trapa natans var. bispinosa*) is rendered infeasible in weedy waterbodies.

1.3.7: Menace To Human Health

Aquatic weeds foster mosquito and snail-borne human and animal diseases. Small, sheltered pools formed between aquatic plants like *E. crassipes, P. stratiotes*, and *Myriophyllum* spp. provide perfect breeding habitats for mosquitoes that are vectors of malaria, dengue fever, encephalitis, and filaria. In drinking waterbodies, certain snails, like *Pila* spp., inhabit aquatic weeds and act as alternate hosts of helminthic worms of human and domestic animals. There are snails which inhabit the roots of waterhyacinth and waterlettuce and act as intermediate hosts of human blood worms like *Bilharzia* spp. Blue-green algae ingested by man with drinking water may cause gastrointestinal disturbances. In livestock, *Microcystis* spp. can cause acute toxicity.

1.3.8: Menace To Aquatic Sport And Aesthetics

Weeds take the charm of water-sports like boating, swimming, and skiing out of waterbodies. They put forth physical hindrance to the water-sport and pollute the water itself. Phytoplanktons turn the water turbid with chlorophylls and the multicellular algae make it scummy. In either case the water becomes unfit for aquatic sport. Besides, the shoreline weeds mar the recreational values of water when they decompose in an unsightly manner.

Aquatic weeds also distract the pleasure of offshore and on shore hotels and residences from large lakes and other waterbodies (Bachmann, *et al.*, 1999). In fact, for any aquatic entertainment only weedfree waterbodies containing clear water, free of unpleasant odour and taste, are considered suitable

SELECTED BIBLIOGRAPHY

1. Akcin, G. *et al.* 1994. Removal of lead by waterhyacinth. J. Envion. Sci. and Health 29 (40): 2177-2183.
2. Bachmann, M. *et al.* 1999. Living At the Lake – A Hnadbook for Florida Lake – Front Property Owners. IFAS Pub. POB-110011, Gainesvilla, FL 32610-0011, 182 p.
3. Batterson, T.R. 1983. An introduction to limnology. Aquatics 5 (1): 16-20.
4. Benson, D.J. and W. Pearson. 1980. Production and energy flow in the benthic community of a texas pond. Hydrobiologia 74: 81-93.
5. Boon, J.J. *et al.* 1983. The analytical pyrolysis of particulate residues of decomposing *Myriophyllum heterophyllum*. Aquat. Bot. 15 (3): 310-320.
6. Boughey, A.S. 1963. The explosive development of a floating weed vegetation on lake Kariba, Adanosonia (n.s.) 3 (1): 49-61.
7. Boyd, C.E. 1972. Sources of CO_2 for nuisance blooms of algae. Weed Sci. 20 (5): 492-497.
8. Brezny, O. *et al.* 1973. Studies on evapotranspiration of some aquatic weeds. Weed Sci. 21 (3): 197-203.
9. Cary, P.R. and K.H. Bowner. 1991. Aquatic plants--friends or foe. Biotrop-Special Pub. No. 40, 7-23 pp.
10. Coquery, M. and P.M. Welbourn. 1994. Mercury uptake from water and sediment by the rooted and submerged aquatic macrophytes. Archives of Env. Contaminant and Toxicology 26 (3): 335-341.
11. Edie, H.H. and B.W.E. Ho. 1969. *Ipomoea aquatica* as a vegetable crop in Hong Kong. Eco. Bot 3 (1): 32-36.

12. Gasith, A. and W. Lawacz. 1976. Breakdown of leaf litter in the littoral zone of eutrophic lake. Ekologia Polska 24: 421-430.

13. George, K. 1976. Conservation of water by *Salvinia* sp. Ind. J. Expt. Biology 14 (3): 379-380.

14. Goulder, R. and D.J. Boatman. 1971. Evidence that N supply influences the distribution of a freshwater macrophyte *Ceratophyllum demersum*. J. Ecol. 59, 783-791.

15. Grasmuch, N. *et al.* 1995. Assessment of the bioindicator capacity of aquatic macrophytes using multivariate analysis. Hydrobiologia 300/301, 115-122.

16. Gupta, O.P. 1987. Aquatic Weed Management. Today and Tomorrow Publishers, N. Delhi (India), 247 p.

17. Hasler, A.D. 1969. Cultural eutrophication is reversible. Biol. Sci. 19:5.

18. Ho, Y.B. 1981. Growth enhancement and evaluation in mineral content of aquatic plants as a result of eutrophication. Ind. J. Biochem. and Biophys. 18 (4).

19. Holm, L.G. *et al.* 1969. Aquatic Weeds. Science (N.Y.) 166, (3906): 699-709.

20. Hunter, R.D. 1976. Changes in the carbon and nitrogen contents during decomposition of three macrophytes in freshwater and marine environments. Hydrobiologia 51: 119-128.

21. Ikusima, I. 1965. Ecological studies on the productivity of aquatic plant communities. Bot. Mang. Tokuo 78: 202-211.

22. Krishnamoorthi, K.P. and G. Visveswarna. 1965. Hydrobiological studies with reference to sudden fish mortality. Hydrobiol. 21: 275-303.

23. Kumar, T.S. *et al.* 1983. Azolla as a source of biomass: a comparison with waterhyacinth (a survey). Int. J. Environ. Studies 20 (3/4): 275-279.

24. Lewis, M.A. 1996. Use of freshwater plants for phytotoxicity testing: a review. Environ. Pollution 81 (3): 319-336.

25. Little E.C.S. 1967. Progress report on transpiration of some tropical plants. PANS (C) 13 (2): 127-132.

26. Maberly, S.C. and D.H.N. Spence. 1983. Photosyntheitc inorganic carbon use of freshwater plants. J. Eco. 71 (3): 705-724.

27. Mackenthum, K.M. 1971. Nutrients and their relationship to weed and algal growths. Hyacinth Contr. J. 9 (1): 58-61.

28. Mahapatra, I.C. *et al.* 1971. Effect of N.P.K. fertilisation on prevalence of algae under field conditions. Ind. J. Agro. 16: 19-22.

29. Mahendranath, T. 1970. Waterhyacinth has value as pig feed. Pig Farmer, Jan., 599-607 pp.

30. Marton, J.F. 1975. Cattail (*Typha* spp.) - Weed problem or a potential crop. Eco. Bot. 28 (1): 7-29.

31. Majid, F.Z. 1999: Aquatic Weeds: Utility and Development. Agro. Botanica, Bikaner (India), 96 p.

32. Mehta, I. and R.K. Sharma. 1976. A note on the water loss from *Typha* (*aira*) weed. Annals. Arid Zone 15 (1&2): 114-116.

33. Moenandir, J. and Murgito, 1994. Heavy metal adsorption by aquatic weeds. Agrivita 17 (2): 61-64.

34. Moore, A.W. 1969. Azolla: biology and agronomic significance. Bot. Rev. 35 (1): 17-34.

35. Murugesan, A.G. and N. Sukumuran. 1999. Potential utilization of aquatic weeds for treating industrial effluents. (In) Limnological Research in India, (Ed.) Daya Pub. House, Delhi, 247-259 pp.

36. Rajnik *et al.* 1996. Mosquito breeding in relation to aquatic vegetation in rice fields of Gujarat. Ind. J. Malariology 33 (1): 30-40.

37. Reedy, K.R. 1984. Nutrient removal potential of aquatic plants. Aquatics 6 (1): 15-16.

38. Roger, H.H. and D.E. Davis. 1972. Nutrient removal by waterhyacinth. Weed sci. 20 (5): 423-428.

39. Satpathy, B. 1964. Kalamisag (*Ipomoea reptans*) - a new addition to our greens. Ind. Fmg. 14 (8): 12, 16.

40. Schnoor, J.L. 1980. A steady state eutrophication model for lakes. Water Res. 14: 1651-1665.

41. Seidal, K. 1971. Macrophytes as functional elements in environment of man. Hydrobiologia 12: 121-130.

42. Sevicham, P.J. and P.V. Madhusoodan 1995. Comparative performance of three species of azolla as a biofertiliser in paddy fields. J. Eco. and Taxonomie Botany 19 (3): 683-686.

43. Sheffield, C.W. 1970. Eutrophication and aquatic weeds. Hyacinth Contr. J. 8 (2): 26-28.

44. Singh, R.N. 1961. Role of Blue-Green Algae in Nitrogen Economy of Indian Agriculture. I.C.A.R., New Delhi, 175 p.

45. Sinha, S.N. and L.P. Kinha. 1969. Studies on the use of waterhyacinth culture in oxidation ponds for treating digested sugar waste and effluent of septic tank. Environ. and Health 11: 197-207.

46. Sinha, S.N. *et al.* 1994. Accumulation of toxicity of iron and manganese in *Spirodela polyrhiza*. Bull. Env. Contamination and Toxicology 53 (4): 610-617.

47. Steward, 1970. Nutrient removal potentials of various aquatic plants. Hyacinth Contr. J. 8 (2): 34-35.

48. Subba Rao N.S. 1984. Biofertilizers in Agriculture. Oxford IBH Publishing Co., New Delhi, 186 p.

49. Sulton, D.L. and W.H. Ornes. 1975. Phosphorus removal from static sewage effluent using duckweed. J. Environ. Quality 4: 367-380.

50. Thomas, E.A. 1983. Aquatic weed–ecology and health aspects–floating and decaying weeds. Water Supply 1 (1): 57-65.

51. Truax, R.E. *et al.* 1972. Duckweed for chick feed. Lousinana Agri. 16 (1): 8-9.

52. Van den Berg, M.S. and H. Coops, 1999. Stoneworts – Valuable for Water Management. Harry Hosper Riza, POB-17, 40 p.

53. Vollenweider, R.A. (Ed.) 1969. A manual on methods of measuring primary production in aquatic environment. IBP Handbook No. 12, Blackwell Sci. Pub., Oxford.

54. Wakefield, J.W. 1962. Effects of water pollution on aquatic vegetation. Hyacinth Control J. 1: 12.

55. Wang, W, and M.A. Lewis. 1997. Metal accumulation by aquatic macrophytes. IN: Plants for Environmental Studies, W. Wang, *et al.* (Eds), CRC Leurs Publishers, N.Y. 367-416 pp.

56. Welch, P.S. 1952. Limnology. Mc Graw Hill, N. York, 538 p.

57. Zranyika, M.F. and T. Ndapwadza. 1995. Uptake of Ni, Zn, Fe, Co, Cr, Pb, Cu, and Cd by waterhyacinth. J. Environ. Sci. and Health. 30 (1): 157-169.

58. Zaranyika, M.F. *et al.* 1994. Uptake of Zn, Co, Fe, and Cr. by waterhyacinth in lake Chivero, Zimbawe. Science of the Total Environment 153 (1-2): 117-121.

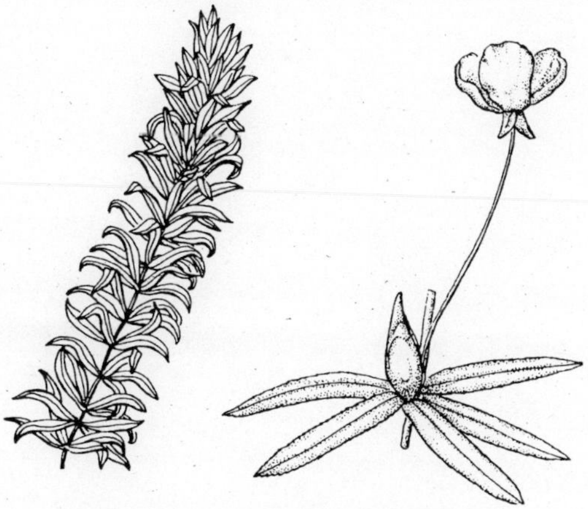

Egeria densa (Waterweed)

Chapter 2

PROPAGATION, DISPERSAL, AND KINDS OF AQUATIC PLANTS

2.1: PROPAGATION AND DISPERSAL

The vascular aquatic plants may propagate by seeds (Fig 2.1), turions, bulbs, and rhizomes (Fig. 2.2). In certain cases even fragments of stems and leaves are capable of developing into new plants and colonies, for instance, *Hydrilla, Myriophyllum, Cabomba*, and *Najas* spp. Seeds and fruits of water-plants are dispersed short or long distances, primarily with wind and water. *Typha* spp. are very common plants that disseminate their fruits fast with wind. Seeds and fruits of several water plants are carried down the stream and are dispersed with water currents. Some of these seeds and fruits possess special floating devices while others float just because of their light weight. Heavy seeds *e.g.* of waterhyacinth (*Eichhornia crassipes*) sink through the water and settle on hydrosoil. These disperse later with back waters or with bottom muds to new sites.

Navigation carries aquatic plants to new continents as weeds, fruits, and plant fragments. Man has carried some very bad alien aquatic weeds across oceans for their attractive flowers. Waterhyacinth is one such notorious plant in India. Also, small aquatic herbs imported as aquarium plants have often escaped as serious weeds, for instance, water fern (*Salvinia molesta*). There could be other venues of dispersal of aquatic weeds. For instance, when weedy waterbodies are connected to the new ones, the weeds find easy way to disperse to new volumes of water. Floods spread floating plants on vast stretches of lowlands. Also, physical removal of aquatic weeds often helps their dispersal to new situations.

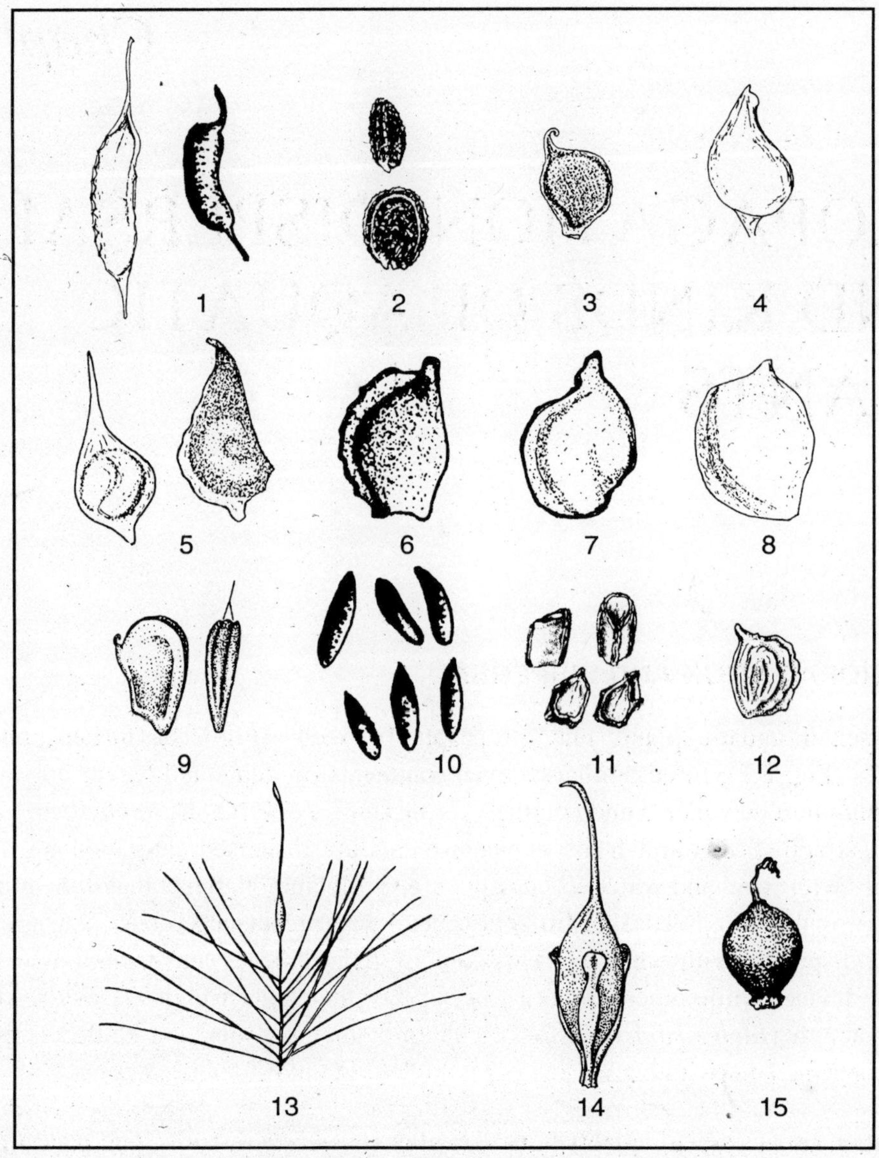

Figure 2.1: Seeds (and fruits) of certain, common aquatic plants. (1) *Zannichellia palustris*, (2) *Nasturtium officinale*, (3) *Ranunculus repens*, (4) *Ruppia maritime*, (5) *Potamogeton crispus*, (6) *P. nodosus*, (7) *P. perfoliatus*, (8) *P. pectinatus*, (9) *Alisma* sp., (10) *Najas* sp. (11) *Jussiea* sp., (12) *Sagittaria* sp., (13) *Typha* sp., (14) *Sparganium* sp., and (15) *Polygonum amphibium*.

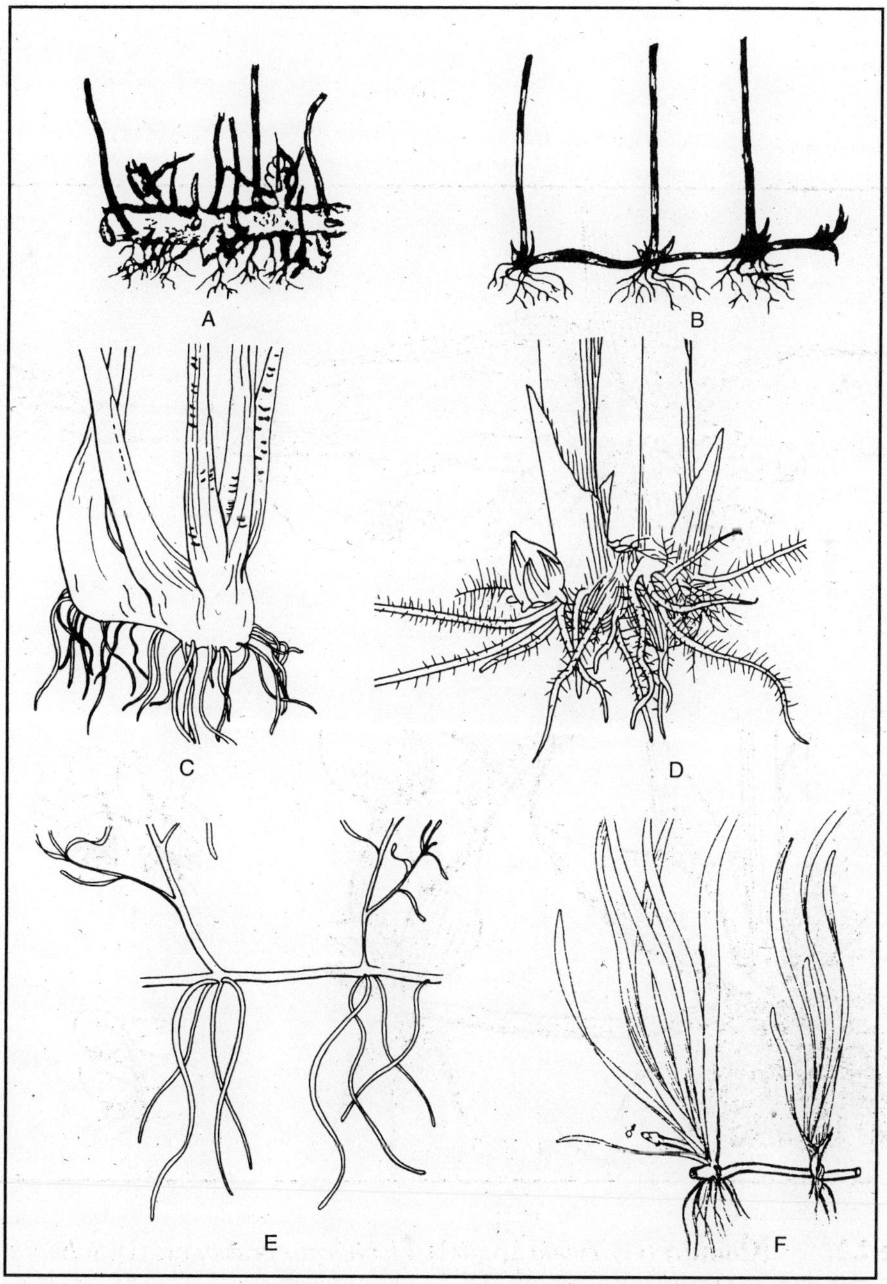

Figure 2.2: Vegetative reproduction propagules of certain aquatic plants (rhizomes, stolons, and offsets), (A) *Nymphaea* sp., (B) *Nelumbo* sp., (C) *Sagittaria* sp., (D) *Scirpus* sp., (E) *Potamogeton* sp., (F) *Vallisneria* sp.

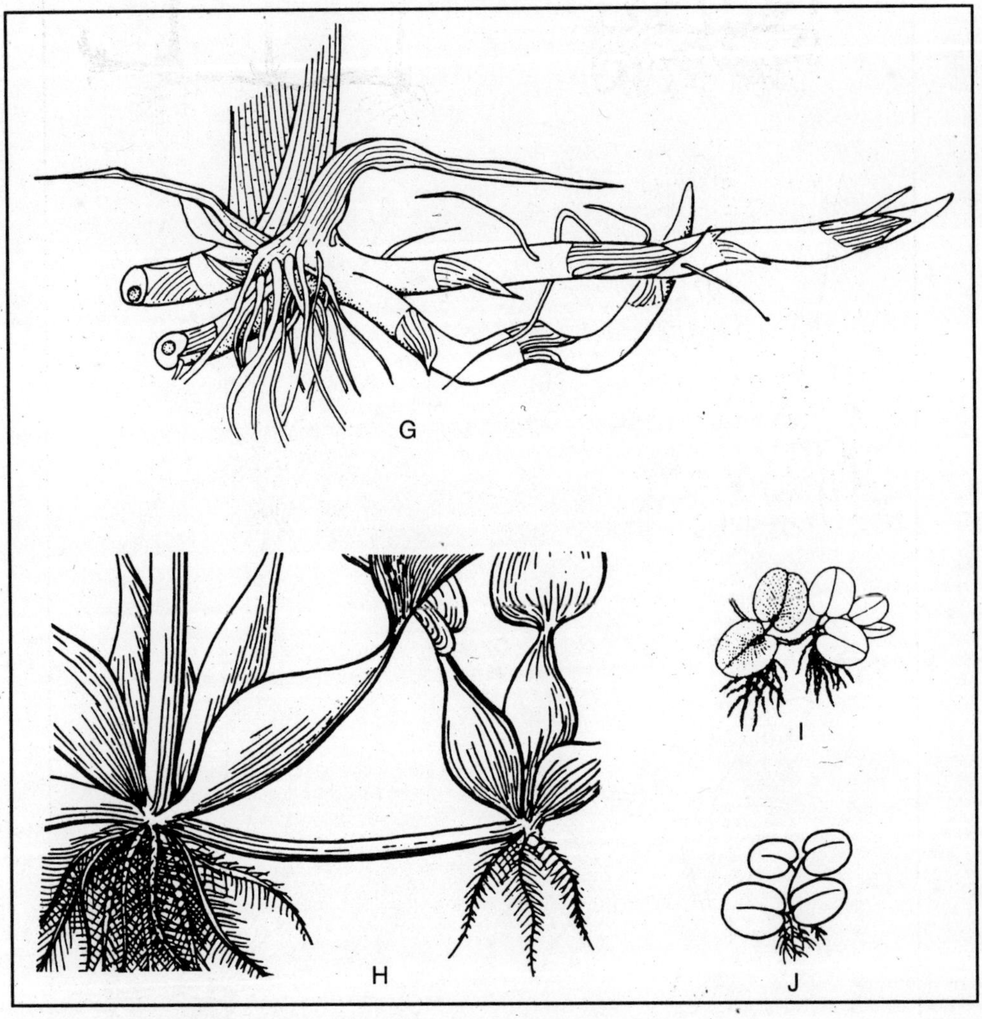

Figure 2.2: **(Contd.)** (G) *Typha* sp., (H) *Eichhornia crassipes*, (I) *Salvinia molesta*, and (J) *S. rotundifolia*.

2.2: KINDS OF AQUATIC PLANTS

If we visualise the cross-section of a large lake, it can be divided from the point of view of aquatic plant growth into three distinct habitats. The first one is the dry to moist, terrestrial periphery called *ditchbank*. It is followed by a gradually increasing depth of water up to usually 3 or 4 m. This is the **littoral** water zone. Beyond this, towards the middle of the waterbody, the water is too deep to support any plant life from its hydrosoil. This is the **liminal** water zone. The initial phase of the littoral water is frequently identified as *shoreline* where the land and water meet. Here the water is very shallow. Each of the above mentioned habitats of a waterbody is usually dominated by distinct aquatic plant species although there may be considerable overlapping in this respect. Therefore, to establish the major habitat of an aquatic plant one must watch it grow for an extended period of time in its natural environment.

Besides the lakes and rivers there are other, special wet situations where specific aquatic vegetation occurs, for example, the marshlands and swamps.

Described below are the major categories or kinds of plants growing in aquatic and semi aquatic environments.

2.2.1: True Aquatic Plants (Water-Plants)

These plants occur directly in the water contained in the littoral and liminal water zones of the waterbodies. These are typical hydrophytes which vary in size from single cells to complex, vascular plants. In the event of drying of the waterbody, the water–plants hibernate in spores, seeds, rhizomes, tubers, and turions. The vascular water-plants, also called macrophytes, are succulent herbs with thin cuticle, a weak vascular system, and lots of aeronchymatous tissue. They absorb dissolved nutrients from their all parts in contact with water. The whole leaf is made-up of mesophyll tissue, with abundant chlorophyll in their epidermal cells.

The water-plants have been further classified into four major biological types*, depending upon their growth in relation to the water surface (Fig. 2.3). These biological types are:-

(a) Surface or floating plants (Biological type-1): The surface plants put forth their foliage flat on the surface of water to form some sort of blanket or mat. These may either be (i) free-floating or (ii) anchored to the hydrosoil with their roots and rhizomes. They are bouyant largely because of their plentiful aeronchymatous tissues in their leaves and

---

* ***Note:*** *Stodola (1967) had considered ten biological types of true aquatic plants. In this text these have been condensed into four types (Fig. 2.3).*

petioles, but sometimes it may also be because of the presence of air trapping hair on their surfaces (*Ex. Salvinia* spp.). Some common free-floating surface plants are *Eichhornia crassipes, Salvinia molesta, Pisitia lanceolata,* and *Azolla* spp. The anchored surface plants include *Nymphaea, Nelumbo*, and *Nymphoides* spp.

It may be interesting to note here that some free-floating plants, upon either recession of water or after getting drifted to nearby shallow ditches and shores, may anchor their roots in soil and lose their floating features.

Certain floating plants may grow into such massive mats as to form sudds which, in turn, may give rise to floating islands. Waterhyacinth and alligatorweed are two such floating weeds with us.

The floating weeds hinder navigation and recreation, elevate water losses, cut the water-flow rates, and reduce light penetration into the waterbody.

(b) Submersed plants (Biological types 2 & 3): The submersed plants grow largely under water though some species may also put forth out-of-water leaves and inflorescence. Such leaves are usually morphologically different from their underwater counter parts (heterophyllous), *e.g. Hydrilla, Myriophyllum*, and *Potamogeton* spp.

Like the surface plants, the submersed plants can be either free-floating (*Ex. Utricularia, Ceratophyllum*, and *Najas* spp.) or anchored to the hydrosoil. (*Ex. Hydrilla verticillata* and *Vallisneria spiralis*). However, even some anchored species like *Hydrilla* may get completely detached from the hydrosoil and continue to grow as apparently free-floating plants within or near the water surface. Such plants can again strike roots whenever the opportunity arises and grow as true submersed, anchored plants. This feature of anchored, submersed plants greatly helps their dispersal to distant sites.

The physiology of submersed species of plants is curiously different from the land plants and some of these features are partially adapted even by the surface and the emersed species. In the first instance, the submersed hydrophytes economise heavily in their use of CO_2 and O_2. Both these gases are present in water in quantities less than 1/30th their content in air. During photosynthesis the submersed plants use, besides the dissolved CO_2, the dissolved bicarbonates. Perhaps, the bicarbonates are enzymatically converted to CO_2 inside the mesophyll cells of leaves with the help of carbonil anhydrase. Also, the oxygen produced during photosynthesis of submersed plants is stored in their special lecunate tissues. These cells are distributed throughout the body of submersed hydrophytes. The stored oxygen is reused by the plants to conduct respiration. It is released by plants only when it is produced in excess of their O_2 hoarding capacity. It then bubbles out the water alongwith other gases like N, CO_2, and CH_4. (Thus, the bubbles released in the *Elodea* experiment showed in the past to students the worldover,

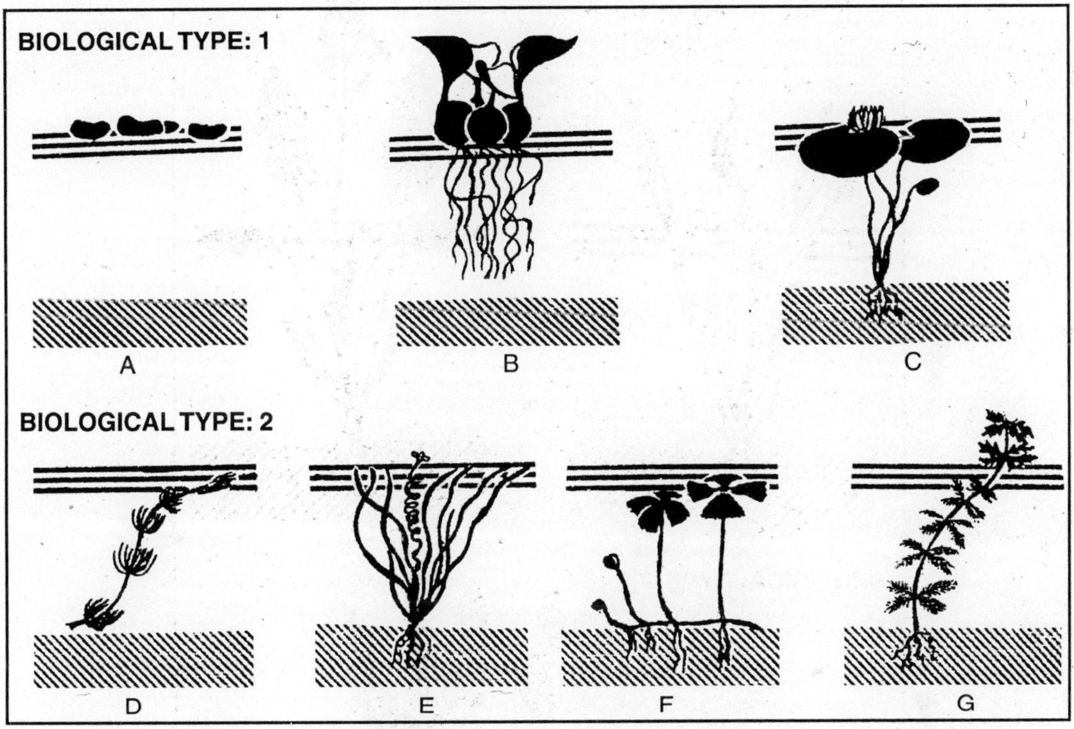

Figure 2.3: Major biological types of true aquatic plants.

Biological type – 1: Surface Floating Plants (A-C). (A) Without roots, (B) With free-floating roots, and (C) With anchored roots.

Biological type –2: Completely Submersed Plants (D-G). (D) Unanchored plants, (E) Anchored plants reaching the water surface, (F) Anchored plants *not* reaching the surface of water, and (G) Anchored plants with some portion of their inflorescence peeping out of water.

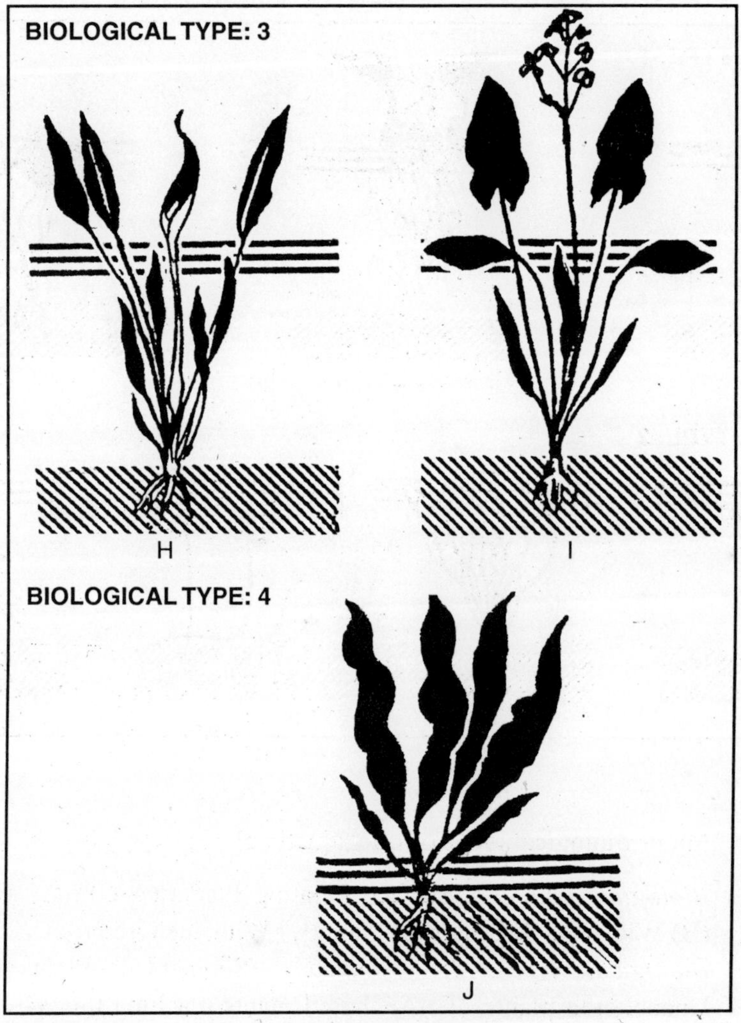

Figure 2.3: **(Contd.)** *Biological type –3* : Partially Submersed Plants (H-I). Anchored plants with their considerable portion above the surface of water. In some plants emersed leaves are different from the submersed ones. Their flowering shoots rise distinctly above the water surface. *Biological type –4* : Emersed Plants (J).

did not actually represent the *start* of photosynthesis). The hydrophytes are also very efficient in the use of limited light that reaches them inside water. Their light compensation point is very low. It is found that of the total solar energy reaching the surface of water only 50% reaches 1 m below it, depending upon the turbidity of water. The light intensity is much less in the lower layers of water. Below 3 m depth the light is so much reduced that usually only a plant like *Hydrilla* can survive and beyond 4 m depth practically no plant life can occur. It is only in very clear waters, which are difficult to find, that the sunlight may penetrate up to 15 m depth. The sunlight under water is cut because of its absorption by water molecules and dissolved and suspended constituents. While the water molecules absorb the ultra-violet and infrared spectrum of light, the organic solutes present in water filter out the blue and violet wavelengths. The suspended mud particles in water absorb all wavelengths of light, almost evenly. Thus, although the intensity of light inside a waterbody is cut greatly, yet it becomes more concentrated in its visible, photo-active spectrum up to a certain water depth.

The submersed weeds cause significant cut in the water flow, encourage seepage, and consequently contribute towards waterlogging and salinity.

(c) Emersed plants (Biological type 4): As the name indicates, these plants tower their major parts of shoots distinctly above the water surface into air and are anchored to the hydrosoil. They usually prefer shallow waterbodies although once established, many of these can withstand considerable, intermittent submergence. This they achieve with the help of their deep and extensive roots and underground stems. *Scirpus, Cladium, Juncus, Alisma, Sagittaria,* and *Cryptocrene* spp. are common emersed plants.

The weedy growth of emersed plants hinder water-flow and choke the shallow waterways, like the drainage ditches.

2.2.2: Shore Plants

The chief characteristic feature of a shoreline is very shallow depth of water near the bank which may sometimes recede to expose its saturated hydrosoil, temporarily. Here, largely semiaquatic plants survive well. Certain shore plants may actually begin their life on moist banks and later creep into the shoreline waters, but mostly they germinate during the period of water recession on the hydrosoil of shore itself. Some typical shoreline plants are: *Typha, Phragmites,* and *Scirpus* spp., *Sparganium erectum, Pontederia cordata, Cyperus iria, Monochoria vaginalis, Marsilea quadrifolia, Potamogeton nodosus*, and *Ipomoea carnea*.

Sometimes a true water–plant may germinate in the shoreline area and later get drifted into the main water. *Eichhornia crassipes* is one such plant. Such plants cannot, of course, be considered as shore plants.

Some of the shore plant species also fill the drainage ditches (see latter)*.

2.2.3: Ditchbank Plants

The banks of various waterbodies, usually called ditchbanks, are also prone to heavy growth of a variety of flora, both short and tall. The low growing grasses like *Cynodon dactylon, Poa* and *Eragrostis* spp,. and *Paspalum notatum* are considered desirable on the banks because of their soil binding and erosion preventing features. But, the tall grasses and all kinds of broadleaf and other plants which hinder approaches to waterbodies, besides hiding rat-holes and soil cracks responsible for bank slides, are considered weedy. In fact, most of the sturdy broadleaf plants having tap roots force heavy cracks in the banks.

The ditchbanks offer typically three moisture zones. In the vicinity of the shores, there is a wet soil zone capable of affording semi-aquatic plant species, particularly the sedges. This is followed in upwardly direction by a moist zone and then a dry zone. The three moisture zones permit the growth of the local terrestrial plants of all kinds, depending upon their moisture responces. There can be considerable overlapping in vegetation through the width of the ditchbanks. Sometimes a weed like *Phragmites karka* growing in the shore waters can also creep onto the banks.

Some common weedy ditchbank plants reported from different states in the country are as under:-

(a) Grasses: *Phragimites communis, P. karka, Sorghum halepense, Apluda mutica, Erianthus ravennae, Narenga prophyrocoma, Rottoboellia exaltata, Saccharum spontaneum, Sorghum halepense, Themeda arundinacea,* and *Vetiveria zizanioides*.

(b) Broadleafs: *Cirsium arvense, Salsola kali, Parthenium hysterophorus, Convolvulus arvensis, Lantana camara, Tribulus terrestris*, and *Glyceria* spp.

(c) Sedges (in the wet zone): *Carex, Scirpus, Cladium*, and *Cyperus* spp.

---

* ***Note:*** *One should differentiate emersed plants from shore plants. The former are true aquatic plants and grow even beyond the shorelines in the main, shallow waterbodies. The shore plants are at most semi-aquatic species though some emersed plants may also be found growing here. It is a positional class of plants in a waterbody rather than of aquatic plants per se.*

2.2.4: Swamp and Marshland Plants

The Oxford dictionary defines swamps and marshes as vast tracts of low lying, wet, spongy lands. But the wet land managers sometimes differentiate between the two; the marshes being under flood for major part of the year and swamps with only seasonal floods.

There is a stagnant mud in a marsh and the water is alkaline to neutral, but not acidic. From the margin of a marsh to its center it shows three kinds of vegetation. In the zone lying nearest to the land, plants with rhizomes and roots adapted to waterlogged conditions *e.g. Scirpus, Carex, Juncus*, and *Equisetum* spp. dominate. In the intermediate zone, anchored plants with tall stems, raised well above the water are dominant, like the *Phragmites* and *Typha* spp. This zone is often called the reed-swamp association zone. The low growing herbs or floating plants are frequently associated with these tall herbs, for instance, *Alisma, Ranunculus, Sagittaria, Ceratophyllum, Hydrilla, Vallisneria, Utricularia*, and *Hydrorrhiza* spp. The inner most zone of marshes carries some typical hydrophytes like *Nymphaea, Nelumbo, Potamogeton, Lemna*, and *Pistia* spp. Often there is certain amount of overlapping in flora of the three zones of marshes.

Our interest in maintaining marshes and swamps is largely for their duck and goose sports. Therefore, sometimes these are also called duck marshes.

SELECTED BIBLIOGRAPHY

1. Biswas, K. and C.C. Calder. 1965. Handbook of Common Water and Marsh Plants of India and Burma. Health Bull. No. 24. Malasia Bureau No. 11, Govt. Press New Delhi, 216 p.

2. Bronmark, C. and L. Hansson. 1998. The Biology of Lakes and Ponds. Oxford University Press. New York, 216 p.

3. Chamberl, P.A. *et al.* 1985. Photocontrol of turion formation by *Potamogeton crispus* L. in the laboratory and natural water. New Phytologist 99 (2): 183-194.

4. Holm, L.G. *et al.* 1977. The worlds' Worst Weeds. East West Centre Book Univ. Press, Hawaii, USA, 609 p.

5. Holzner, W. and M. Numata (ed.). 1982. Biology and Ecology of Weeds--Aquatic Weeds. Centre for Agrobiol. Research, Netherlands, 449-456 pp.

6. Kadono, Y. 1982. Germination of turions of *Potamogeton crispus* L. Physiol. Ecol. Japan 19: 1-5.

7. Menon, M.V.K. 1971. The threatening weeds. Sci. Today 6 (2): 33-38.

8. Pieterse A.H. 1981. A study on certain aspects of seed germination and growth on *Pistia stratiotes* L. Acta. Bot. Neerl. 30 (1/2): 45-57.

9. Richard, J.H. 1982. Development potential of axillary buds of waterhyacinth (*Eichhornia crassipes* Solms). Amer. J. Bot. 69 (4): 615-622.

10. Sastroutomo, S.S. 1981. Turion formation, dormancy, and germination of curly pondweed, *Potamogeton crispus* L. Aquat. Bot. 10: 161-173.

11. Shiam, R. *et al.* 1981. Tuber formation in *Scirpus grossus*. J. Sci. Res. 3 (3): 181-183.

12. Singh, S.P. and R. Sahai. 1974. Effect of certain abiotic factors on the development of root tubers of *Nymphoides indica* in a natural lake. Geobios 1 (6): 181-182.

13. Steward, K.K. 1984. Growth of hydrilla in hydrosoils of different composition. Weed Sci. 32: 371-375.

14. Sotdola, J. 1967. Encyclopedia of Water Plants T.F.H. Pub. Inc., N.J. USA, 368 p.

15. Westlake, D.F., J. Kvet, and A. Szczepanski. 1998. The Production Ecology of Wetlands: The IBP Synthesis. Cambridge University Press, N.Y. USA, 568 p.

16. Winston, R.D. and R.R. Corham. 1980. Turions and dormancy states in *Utricularia vulgaris*. Can. J. Bot. 57 (24): 2740-2749.

17. Yeo, R.R. 1965. Life history of sago pondweed. Weeds 13: 314-321.

18. Yeo, R.R. 1966. Yields of propagules of certain aquatic plants. Weeds 14: 110-113.

Pontederia cordata (Pickernel weed)

Chapter 3

COMMON AQUATIC PLANTS AND THEIR DESCRIPTIONS

Although in India a wide spectrum of aquatic flora is found but in this chapter only the species which have overgrown to the extent of becoming weedy and are imparting imminent effects on our aquatic environment, with resultant economic and/or aesthetic damages, are included. These comprise the weedy algae, ferns, and the angiosperms. No attempt is made here to provide identification keys to the plant species but the weedy plants have been described adequately, alongwith their sketches, to the extent possible.

3.1: ALGAE (THALLOPHYTA)

Plants largely microscopic; thalloid; without distinct roots, stems, or leaves; non-vascular.

3.1.1: Waterbloom Forming Algae

There are three major waterbloom forming algae playing havoc in pisciculture, particularly during the summers (Chapter-1). These algae are:

1. ***Microcystis aeruginosa* Kuntz.** (*Family*: Myxophyceae) **Fig. 3.1(A)**

   It is a colonial, blue-green alga with spherical cells.

2. ***Anabaena spiroides* Klebahn** (*Family*: Nostocaceae) **Fig. 3.1(B)**

   It is a filamentous, blue-green alga with spherical cells.

3. ***Spirulina subsalsa (****Family*: Myxophyceae/Cyanophyceae)

   Cells spheric or cylindric, periphyton.

3.1.2: Scum Forming Algae

These algae, as the name suggests, form ugly scums, floating largely on the surface of water but sometimes under water. Scums are undesirable in several ways (Chapter-1). There are six major scum forming algae recorded in India.

1. ***Spirogyra* sp.** (pond silk) **Fig. 3.1(C)**
   Family: Chlorophyceae (Zegnemaceae)

   It comprises unbranched, grass-green filaments, with chloroplasts in the form of spiral ribbons, up to 15 in number in each cell. Chromatophores present, sexual reproduction by conjugation. Spirogyra occurs in tangled masses, usually floating on the surface of water as scum whenever competition from other floating weeds is non-existent. It also forms partly submerged mats. *Spirogyra* gives a slippery touch.

2. ***Ulothrix* sp.** **Fig. 3.1(D)**
   Family: Chlorophyceae

   It is an unbranched, thread-like, rather fine, cottony scum forming alga. It is anchored to the stones, sticks, and like objects present in the waterbody. Ulothrix threads are not netty.

3. ***Rhizoclonium* sp.** (*kai*)
   Family: Chlorophyceae (Cladophoraceae)

   It is a coarse, unbranched alga with hair-thick, wiry filaments; cell walls thick, lamellate, chloroplasts arranged in parietal, net-like fashion, devoid of conjugation. It forms tangled masses at places where other floating weeds are cleared.

4. ***Hydrodictyon* sp.** (water net) **Fig. 3.1(E)**
   Family: Chlorophyceae (Hydrodictyaceae)

   It is made-up of up to 30 cm long strands (threads), forming hollow, cylindrical network closed at either end, visible to the naked eye. It is capable of trapping fish to death. This alga is more common in still waterbodies than in the flowing ones.

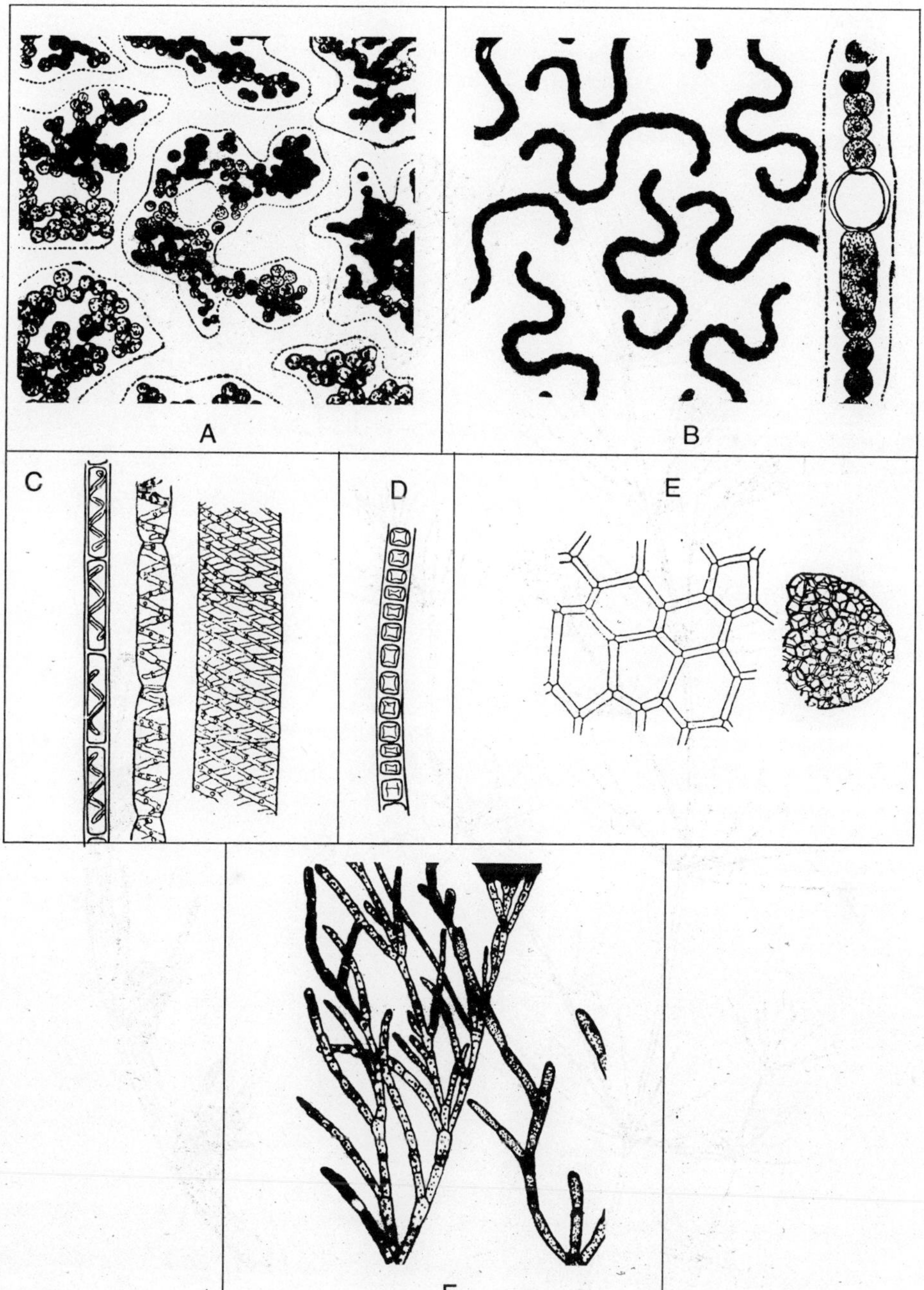

Figure 3.1: Common, weedy, lower algae

A = *Microcystis aeruginosa* Kuntz., B = *Anabaena spiroides* Klebahn, C = *Spirogyra* sp., D = *Ulothrix* sp., E = *Hydrodictyon* sp., and F = *Cladophora* sp.

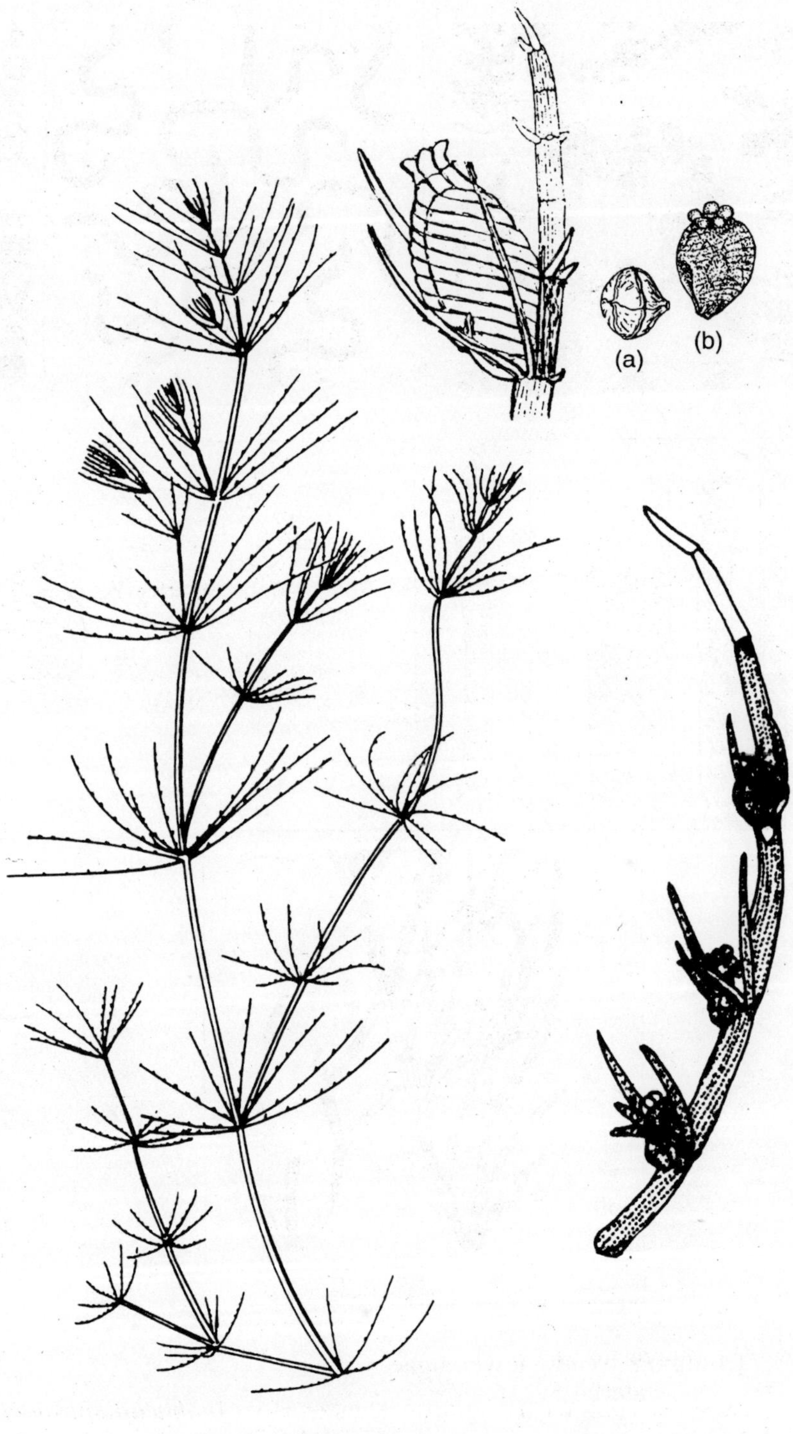

Figure 3.2: *Chara vulgaris* L. (stonewort) (Biotype –2)

5. ***Cladophora* sp.** **Fig. 3.1(F)**
Family: Chlorophyceae (Cladophoraceae)

It is a somewhat branched, filamentous, brush scum forming alga, apparently resembling *Rhizoclonium* sp. It is usually periphyton.

6. ***Pithophora* sp.** (wet wool):
Family: Chlorophyceae (Cladophoraceae)

It is an irregularly branched, thread-like, rather coarse, filamentous alga, initially periphyton but later floating as mats and appearing as masses of wet wool.

3.1.3: Higher Algae

***Chara vulgaris* L.** (stonewort, muskgrass) **(Fig. 3.2)**
Family: Chlorophyceae (Characeae)

It is a macroscopic green alga with thallus bearing nodes and internodes which bear sex organs; sexual reproduction is oogamous. Oogonium is capped with five pointed cells placed above the smaller-sized antheridium in the branch axils. The false stem comprises single cell joints, with cylindrical, whorled branches.

It is a submersed alga, anchored to hydrosoil with its multicellular rhizoid. It falsely resembles a vascular aquatic plant. In sweet waters it is soft to touch but in hard waters it is rough and harsh due to encrustation with lime. When crushed, it emits musk-like odour. A weed of largely lowland paddy fields, although it can invade even canals. Another specie, *C. zelanica* is also reported weedy from some places.

***Nitella hyalina* L.** **(Fig. 3.3)**
Family: Chlorophyceae (Characeae)

It resembles much the *Chara* sp., but its oogonium is capped with 10 cells. It is a smooth alga and it does not emit any musk-like odour.

3.2 FERNS (PTERIDOPHYTA)

1. ***Salvinia molesta* Mitchell** (African payle, floatmoss, water-fern, velvetweed, karibaweed, giant salvinia, *kiambang*) **(Fig. 3.4)**
Family: Salviniaceae

A non-flowering, free-floating, perennial fern, found weedy in all kinds of waterbodies and lowland paddy fields, particularly in the state of Kerala.

Figure 3.3: *Nitella hyalina* L. (stonewort) (Biotype –2)

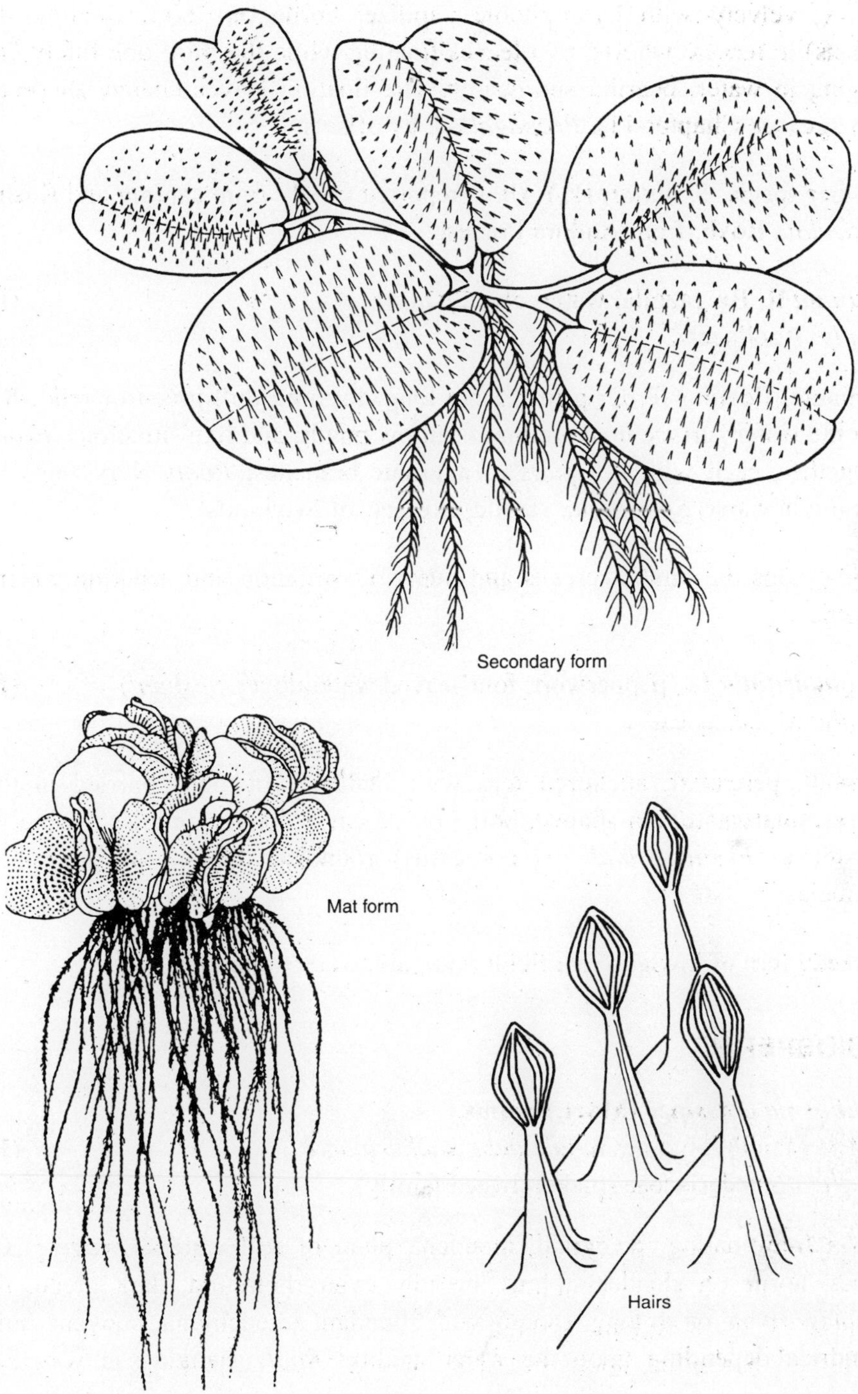

Figure 3.4: *Salvinia molesta* (waterfern) (Biotype-1)

Leaves, velvety with hydrophobic papillae, borne on short, floating rhizomes (offsets) in ternate whorls, two leaves floating while the third one finely dissected, hanging in water, bearing sporocarps. The floating leaves change shape and size with age (see Chapter-11). *Propagation*, by offsets.

Another specie, *S. natans* (L.) All. is reported weedy from Jammu and Kashmir and *S. cuculata* Roxb. from Andhra Pradesh.*

***Azolla pinnata* R. Br.** (azolla, watervelvet, *pana*) **(Fig. 3.5)**
Family: Azollaceae

A minute, floating fern, up to 2.5 cm long, forming dull green or reddish blanket over the water surface in shorelines, shallow ponds, and like situations. *Fronds*, 1 to 3 together, each with 3-5 pairs of alternate branches; *Roots*, very fine, plumose, hanging in water; *Sporangia,* sessile, axillary, of two kinds.

Azolla clogs the intake screens and pums of irrigation and drinking water supply systems.

***Marsilea quadrifolia* L.** (pepperwort, four-leaved waterclover, *sushuni*) **(Fig. 3.6)**
Family: Marsileaceae

A small, perennial, anchored fern with slender rootstocks burried in the mud, *Leaves*, quaternate, fan-shaped, borne on 15 cm or longer petioles arising from the rootstocks; *Fruiting bodies* (sporocarps) roundish, black, on short branching peduncles.

A weedy fern of lowland rice fields and quiet water situations. *

3.3: ANGIOSPERMS

1. ***Eichhornia crassipes* (Mart.) Solms**
(waterhyacinth, blue-devil, *jalkumbi, kachuripana*) **(Fig. 3.7)**
Family: Pontederiaceae (pickerelweed family)

It is a free-floating, perennial, monocot, plant of still waters. *Leaves*, 10-15 cm across, borne on slender stolons, broadly ovate-rhomboid, dark green, rounded; *Petioles,* 30 cm or so long, spongy with abundant aerenchyma, buoyant, bulbous or cylindrical depending upon the water quality; *Roots*, hanging in water, fibrous,

---

Note: *1. Thallus = The flat, green body of some lower plants.*
2. Frond = a green, leafless stem; falsely appearing leaf-like.

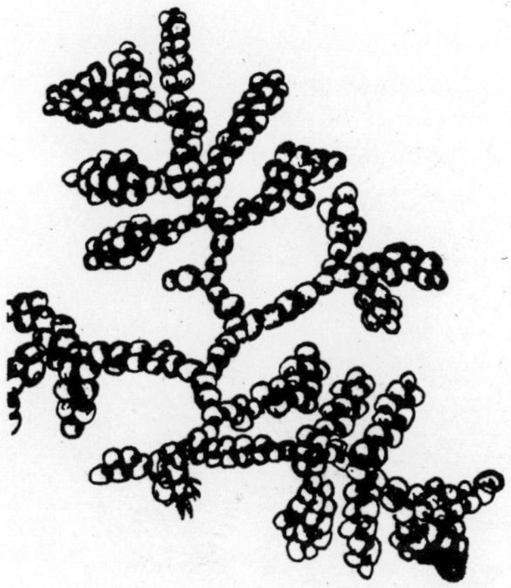

Figure 3.5: *Azolla pinnata* R. Br. (watervelvet) (Biotype-1)

Figure 3.6: *Marsilea quadrifolia* L. (pepperwort) (Biotype-2)

Figure 3.7: *Eichhornia crassipes* (Mart.) Solms (waterhyacinth) (Biotype-1)

emerging from numerous stems, sometimes anchored in the silted shores, drainage ditches, and channels; *Inflorescence*, a terminal, gorgeous spike, 15-30 cm high; *Flowers*, funnel-shaped, about 5 cm across, mauve or lilac with yellow patch at the centre; *Fruit*, a capsule; *Seeds*, viable, oval at the base and tapering above, ridged; *Propagation*, by offsets and seeds.

Waterhyacinth is one of the most dangerous weeds invading all kinds of waterbodies including lakes, reservoirs, streams, canals, rivers, channels, and even lowland paddies.

2. ***Pistia stratiotes* L.** (waterlettuce, tropical duckweed, *kumbi, pana, akasthamri, takapana)* **(Fig. 3.8)**
Family: Araceae (arun family)

A free-floating, stoloniferous, monocot plant with a tuft of long, fibrous roots beneath. *Leaves*, in decorate rosettes, pale green, sessile, broadly obovate with a truncate, rounded or broadly emarginate apex, up to 13 cm long or 5 cm wide, tomentose near the base, with 7-15 nerves prominent below; *Inflorescence*, a very small spidex located within an up to 10 mm spathe; *Flowers*, minute; *Fruit*, with numerous tiny seeds.

A serious weed of variable waterbodies, lowland paddy fields, and back-water areas.

3. ***Cryptocryne retrospiralis* (Roxb.) Fisher ex. Wydler** **(Fig. 3.9)**
Family: Araceae (arun family)

A rooted, perennial, emersed, monocot amphibious herb from *rootstocks*. *Leaves*, linear, up to 40 cm long, leathery; *Inflorescence*, an ornamental spadix, about 12 cm long, placed in a spiral spathe, red or purple inside, emersed.

The plant is found weedy in lowland paddies, swamps, tanks, and like situations.

4. ***Lagenandra ovata* (L.) Thaw.** *(rukh-alu)* (=*L. toxicaria* Dalzell) **(Fig. 3.10)**
Family: Araceae (arun family)

An anchored, perennial, emersed, monocot, amphibious herb with thick, poisonous rootstocks which trail in the wet mud. *Leaves*, elliptical, leathery, 20-24 cm long and 7-12 cm wide, with pointed bracts and 45-60 cm long petioles growing from the rootstocks; *Inflorescence*, a spadix placed in above half of a spiral spathe. It is found weedy in lowland paddy fields, swamps, tanks, and like aquatic situations.

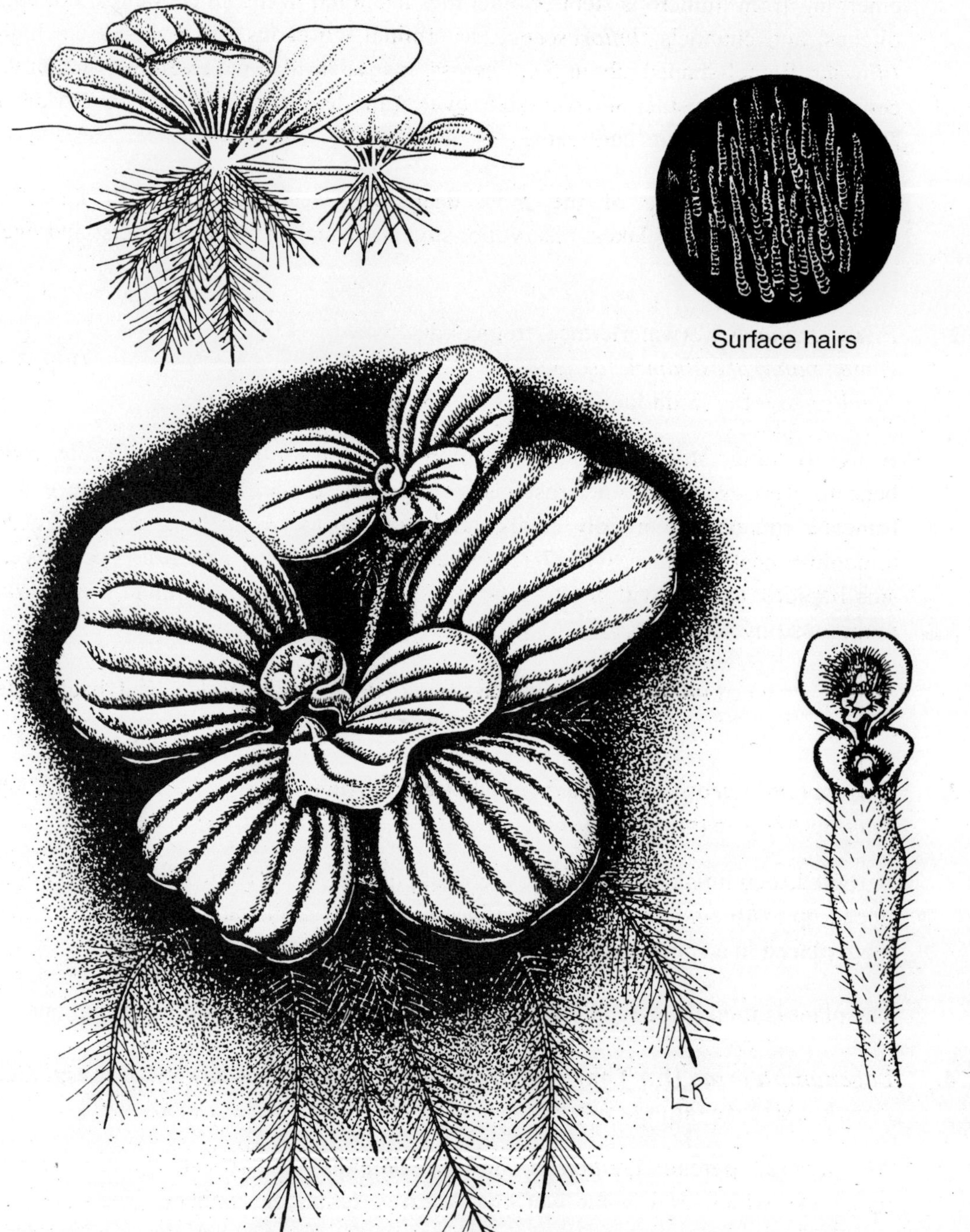

Figure 3.8: *Pistia stratiotes L.* (waterlettuce) (Biotype-1)

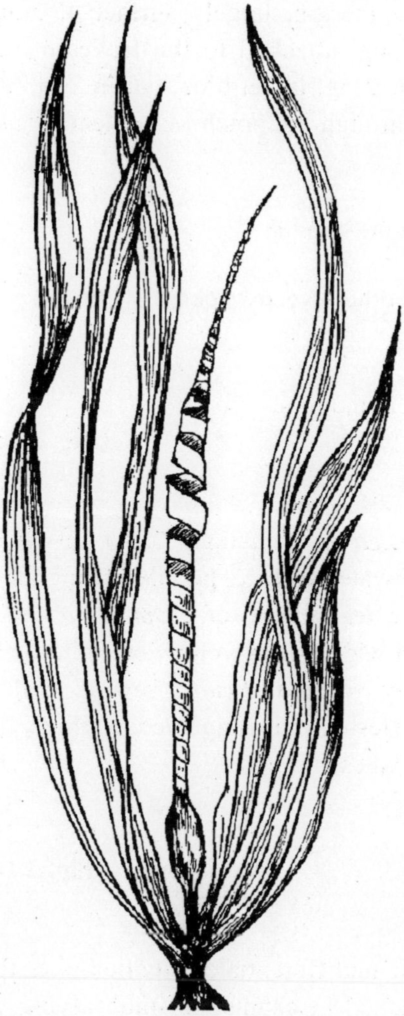

Figure 3.9: *Cryptocryne retrospiralis* (oxb.) Fisher ex. Wydler (Biotype-3)

Figure 3.10: *Lagenandra ovata* (L.) Thaw. (Biotype-3)

5. ***Nymphaea stellata* Willd.** (water-lily, pond-lily, *neelkamal*) **(Fig. 3.11)**
Family: Nymphaeceae (water-lily family)

A perennial, dicot, anchored, submersed plant arising from thick, fleshy rootstocks with tuberous off--shoots which detach to give rise to new plants. *Leaves*, oval to round, 10-20 cm wide, with a deep slit in the margin, largely surface floating though sometimes slightly elevated; *Petioles*, long, attached to the leaves at the summit of their deep notch; *Flowers*, solitary, floating, large, blue, purple or pink, showy, from rootstocks (scapigerous), peeping through the notch of the leaf; *Fruit*, a leathery berry; *Seeds*, arillate.

It has become weedy in many lakes and ponds in the country.

N. nouchali Burm. f. (*Kanval*) **Fig. 3.12** is another weedy specie with sinvate-toothed leaves and white or red flowers.

6. ***Nelumbo nucifera* Gaertn** (sacred lotus, water-chinquapin, scented lotus, *Kamal, padma*) **(Fig. 3.13)**
Family: Nymphaeaceae (=Nelumbonaceae)

A perennial, dicot, anchored, submersed plant. *Leaves*, orbicular, 20-40 cm wide, peltate, entire, elevated above or floating on the water surface, petiole arising from the rhizomes and attached at the middle of the leaf blade (cf. *Nymphaea* spp.); *Flowers*, scapigerous, solitary, elegant, 12-25 cm wide, sulfur-yellow or white-pink, elevated well above the water surface, sweetly scented; *Fruits*, several nutlets placed in a spongy receptacle; *Propagation*, by fleshy, multi-tunneled, trailing, 1-3 cm thick rhizomes and seeds. A weed of largely lakes and ponds.*

7. ***Lemnids*** (duck's meat, water lentils, sea mosses, water-weeds, duck-weeds) **(Fig. 3.14)**
Family: Lemnaceae

Free-floating, tiny, monocot plants with almost undifferentiated, thallus-like, flat bodies; *Thalli* solitary or in groups of 2-6 attached in chains and multiplying by budding to form green carpet on the water surface; *Inflorescence*, in a membranous, minute spathe with single female and two male flowers; *Fruit*, an urticle; *Propagation*, by seeds and frond division.

---

* ***Note:*** *Spathe = An involucral leaf, almost enwrapping the flowers or their bases.*
Spadix = A fleshy spike (of flowers)

Figure 3.11: *Nymphaea stellata* Willd. (water-lily) (Biotype-1)

Figure 3.12: *Nymphaea nouchali* Burm. F. (Biotype-1)

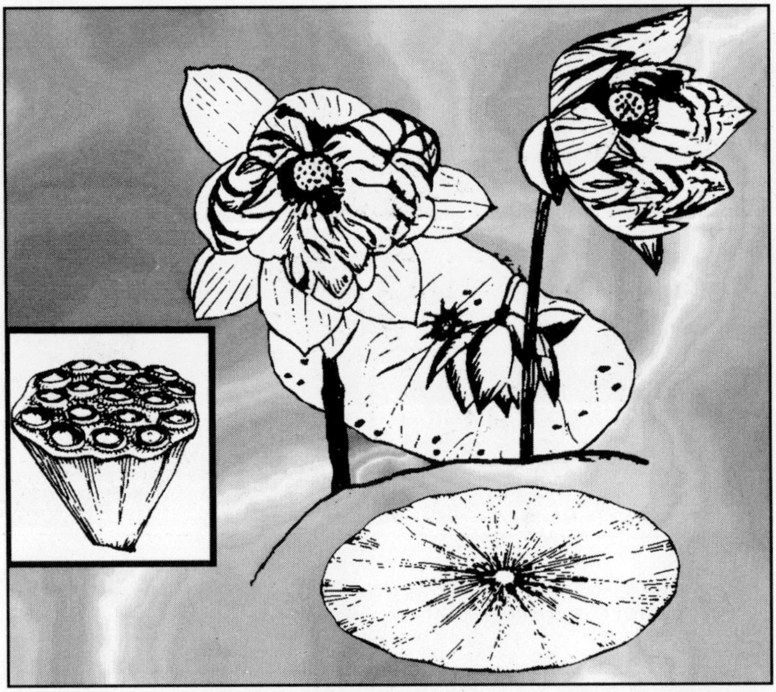

Figure 3.13: *Nelumbo nucifera* Gaertn (sacred lotus) (Biotype-1)

The lemnids form dense carpets to appear as scums in still waterbodies. They become problematic when the infested water is pumped either for irrigation or for supplying drinking water whence these choke the intake screens and pumps. Similar problem can be posed by these tiny plants in pesticide spray systems when lemnid-infested water is used for pesticide application.

7.1 Lemna perpusilla J. Torrey (duckweed) Fig. 3.14(A)

Fronds, almost oblong, up to 5 mm long, with 1-3 nerves and hook-shaped papilla at the tip; *Root*, present, solitary, hanging in water; *Spathe*, one on either side of axis; *Joints* (of plants) sessile, asymmetrical with unequally rounded sides; *Carpet*, green, both above and beneath..

7.2 L. minor L. (lesser duckweed, chlota-pana) Fig. 3.14(B)

It differs from *L. perpusilla* in bearing largely symmetrical plant joints.

7.3 L. trisulca (star duckweed, indrakaripana) Fig. 3.14(C)

Its fronds are oblanceolate and serrated.

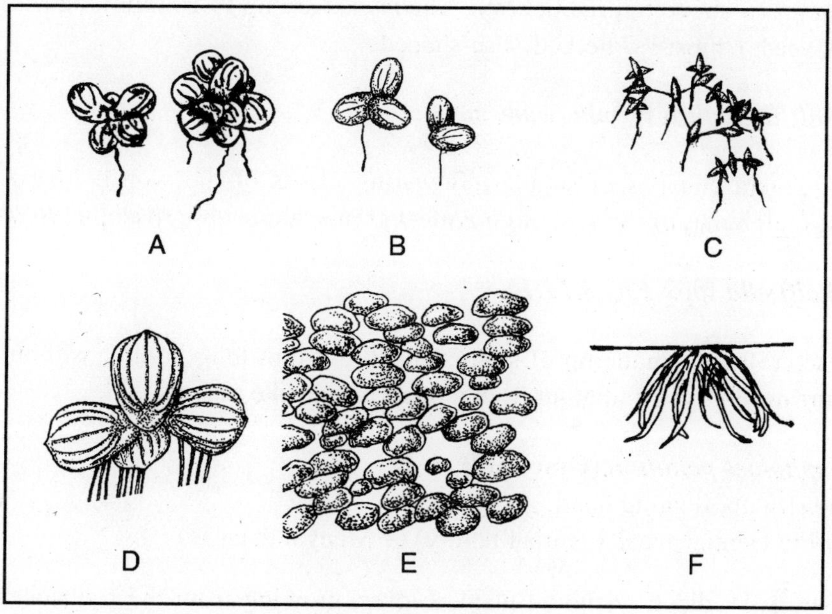

Figure 3.14: Lemnids (water-lentils) (Biotype-1)

A = *Lemna perpusilla* J. Torrey (duckweed), B = *L. minor* (lesser duckweed), C = *L. trisulca* (star duckweed), D = *Spirodela polyrhiza* (L.) Schleid. (giant duckweed), E = *Wolffia arrhiza* Wimm (watermeal), and F = *Wolffiella* spp.

7.4 *L. gibba (common duckweed):*

Its fronds are gibbous, opaque.*

7.5 *Spirodela polyrhiza (L.) Schleid. (giant duckweed, big duckweed) Fig.3.14(D)*

This genera differs from *Lemna* in possessing several roots with each joint and in being green above but red beneath. The *fronds* are up to 10 mm long (largest of the duckweeds), mostly 7 nerved, disc-shaped.

7.6 *Wolffia arrhiza Wimm (watermeal):*

This genera consists of smallest flowering plants of the world. *Fronds*, 1-1.5 m across, globular, of the size of corn meal grains, almost microscopic; *Roots*, absent.

7.7 *Wolffiella spp.: Fig. 3.14(F)*

These are plants made up of strap-shaped, 6-8 mm long bodies, without roots and occurring singly or radiating from a point in star-like masses.

8. *Nymphoides peltatum* (Gmelin) B. & R.
(waterfringe, floating heart, *goolga*) **(Fig. 3.15)**
Family: Gentianaceae (gentian family) (= Menyanthaceae)

A perennial, dicot, anchored plant. *Leaves*, growing from the rootstocks, petioled, blades leathery, heart-shaped or round, dotted on the underside, 15 cm across with smooth margins, the older ones dentate; *Flowers*, on long stalks in hollows, golden yellow, funnel-shaped; *Fruit*, an egg-shaped capsule; *Propagation*, by rootstocks, leaves, and new shoots.

This plant is found weedy in rice fields, muddy pools, slow flowing waters, and periodically flooded areas. Other weedy members are *N. indica* (L.) O.K. and *N. crista* (Roxb.).

---

* **Note:** *1. Scape = A leafless or nearly so stem bearing flowers and arising directly from roots or rhizomes.*

2. Thallus = A flat, green undifferentiated body of some plants.

3. Frond = A green, leafless stem.

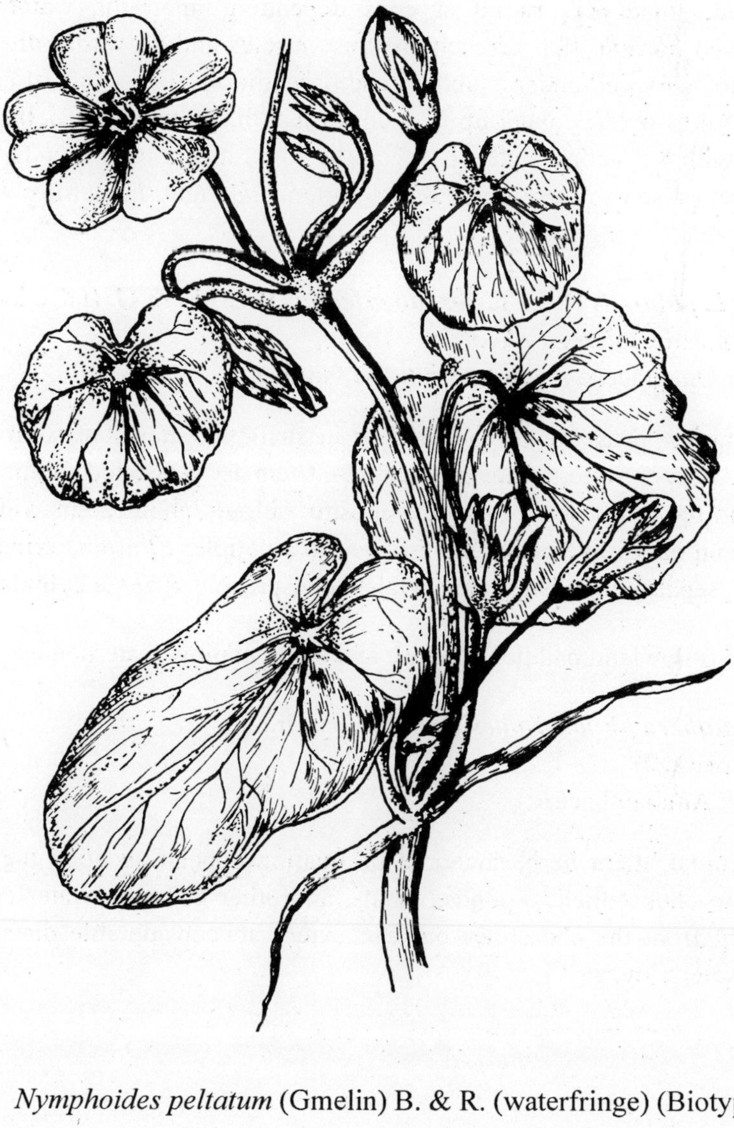

Figure 3.15: *Nymphoides peltatum* (Gmelin) B. & R. (waterfringe) (Biotype-1)

9. *Jussiea repens* Linn. var. *glabrescens* Kuntz. **(Fig. 3.16)**
(= *Ludwigia adscendens* (L.) Hara) (waterprimrose, creeping waterprimrose, primrose, willow, *labanga-phool*)
Family: Onagraceae (evening-primose family)

A perennial, succulent, dicot, creeping or floating weedy herb of marshes,lowland paddies, ponds, and ditches. *Stems*, thick, rooting at lower and middle nodes, branched, sometimes raised at ends depending upon the water depth; lower, submersed portion with aerenchymatous vesicles and upwardly directed breathing roots to survive under submergence conditions; *Leaves*, alternate, oblong, lanceolate to oblanceolate, up to 9 cm in length, prominently petiolated; *Flowers*, showy with 4-5 yellow petals borne at the summit of a long, slender ovary, pedicel and bract present; *Fruit*, a cylindrical pod; *Seeds*, light, numerous; *Propagation* by stolons and floating seeds. [1]

10. *Ludwigia palustris* (L.) Ell. *var. americana* (D.C.) F. & G. (false loosestrife, waterpurslane, *labanga-phool*)
Family: Onagraceae (evening-primrose family)

Plants nearly prostrate on the mud or in shallow waters but sometimes raised at ends in deeper waters; *Stems*, thick 15-30 cm long, jointed, branched, rooting at nodes in upper portions; *Leaves*, opposite, oblong, about 2 cm wide, bright green above and dark olive beneath, narrowed to a petiole; *Flowers*, axillary, small, bell-shaped, sepals yellowish-green, petals inconspicuous *Fruit*, a cylindrical pod.

A weed of lowland paddies, ditches, and other shallow waterbodies.

11. *Alternanthera philoxeroides* (Mart.) Griseb.
(alligatorweed) **(Fig. 3.17)**
Family: Amaranthaceae

A perennial, dicot herb, anchored or floating depending upon the situation. It is found to choke ditches, ponds, canals, and other common water courses. It starts growing from the shorelines but can extend to considerable distances inside the main waterbody.

---

[1] ***Note**: Some authors have dealt Jussiaea under Ludwigia*

Figure 3.16: *Jussiea repens* Linn. *Var. glabrescens* Kuntz. (waterprimrose) (Biotype-2)

Figure 3.17: *Alternanthera philoxeroides* (Mart.) Griseb. (alligatorweed) (Biotype-2)

Stems, branched, prostrate or decumbent, freely creeping but ascending at the edges, up to 50 cm in length, upper internodes slightly pubescent, lower ones rooting at nodes; *Leaves*, opposite, leafblades somewhat fleshy, linear to linear oblanceolate, 2-10 cm long, narrowed to the base into prominent petiole shapes; *Flowers*, in round spike heads borne on long peduncles, silvery white; *Fruit*, a flattened utricle; *Seeds*, lens-shaped, smooth; *Propagation*, vegetative from the rooting nodes and by seeds.

12. *A. sessilis* (Linn.) R. Br. ex DC.

This specie apparently differs in possessing shorter, only up to 4 cm long leaf--blades, elliptic or elliptic-obovate or linear-oblong in shape, and *not* narrowing down to petiole shapes. The flower heads are sessile.

Its habitat is the same as of *A. philoxeroides*.

13. *Polygonum amphibium* L. (water-smartweed, water-knotweed)
(= *P. natans* Eat.) **(Fig. 3.18)**
Family: Polygonaceae (buckwheat family)

A perennial, amphibious, dicot herb, reproducing by seeds and slender, tough forking rhizomes, stolons, and rooting stems. *Stems*, branched, prostrate, trailing through the water or sprawling on the mud, the young stems with tubular stipules, rooting freely at nodes, tending to ascend above the water surface at the apical ends when needed; *Leaves*, alternate, elliptic-oval to lanceolate, petioled, with cylindric stipules; *Inflorescence*, a cylindric, oval, or egg-shaped, 1-4 cm long, axillary spike borne on erect, glabrous, long peduncles; *Flowers*, pink or rose; *Roots*, fibrous, hanging in water from nodes, but anchored in shallow waters.

The plant is found weedy in quiet waters and drainage ditches.

14. *P. glabrum* Willd.

An erect or decumbent, stout, annual herb, *Stems*, 60-150 cm high; *Leaves*, alternate, up to 17.5 cm long, petioled, lanceolate, acuminate with tuberous stipules, *Inflorescence*, branched, erect, cylindric, tapering spikes borne on penducles; *Flowers,* pink.

15. *P. plebeium* R. Br.

A prostrate, much branched, perennial herb. *Stems*, slender, radiating from the top of the woody main roots, spreading; *Leaves*, small, up to 2.5 cm long, alternate, usually oblong or narrow, petioles short, with tubuler stipules surrounding the stem;

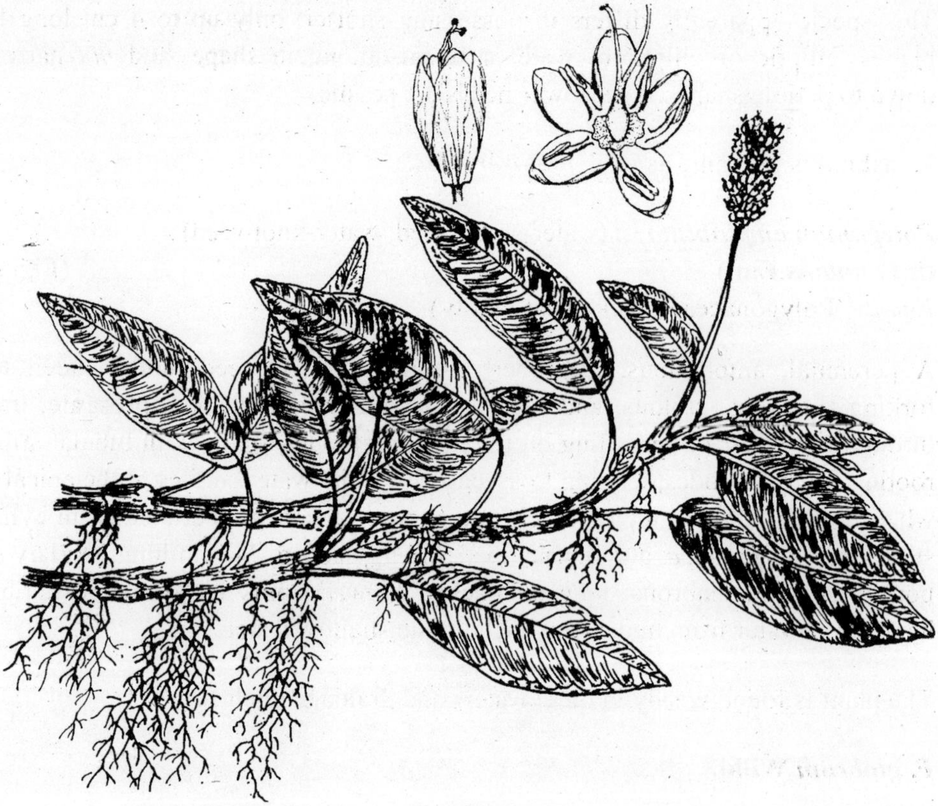

Figure 3.18: *Polygonum amphibium* L. (water-smartweed) (Biotype-2)

Flowers, in axillary clusters of 3 or more, small, pink, shortly stacked; *Fruit*, a beaked nut.

16. ***Nasturtium officinale*** R. Br.
(watercress, awlwort, *piryea, halim*) **(Fig. 3.19)**
Family: Cruciferae (cress family)

A perennial, dicot, marsh or aquatic herb, reproducing by seeds and rooted stems. *Stems*, succulent, much branched, creeping or floating, rooting amply at nodes, ascending in their floating form; *Leaves*, pinnate, leaflets roundish to oblong or elongate, somewhat fleshy, the terminal ones roundish; *Flowers*, in racemes, white; *Fruit*, a linear-cylindric pod with 2 rows of seeds.

It is a weed of ponds and ditches.

17. ***Ipomoea aquatica*** **Forsk.** (water spinach, swamp morningglory, *ganthian, nali*) **(=*I. reptans* Poiret)** **(Fig. 3.20)**
Family: Convolvulaceae (bindweed family)

A perennial, dicot, amphibious herb, floating in water alongwith its hanging roots along the shores but trailing as an anchored plant in wet muds. *Stems*, profusely branched, sub-succulent, slender, hollow, equipped with white, spongy floats, rooting amply at nodes; *Leaves*, elliptic or ovate-oblong; *Flowers*, white or pink, bell-shaped; *Propagation*, by seeds and stem fragments which root at nodes.

The vegetative propagation of this herb leads to fast formation of network which, in association with waterhyacinth, tends to form floating islands.

A weed of ponds, lakes, and canals.

18. ***I. carnea*** **Jacq.** (*kalmisag*, gramophone flower) **(Fig. 3.21)**
(= *I. fistulosa* Mart. Ex. Choisy)
(= I. carnea sub* sp. *fistulosa)

A perennial, dicot, diffuse or straggling shrub of ditches and waterlogged situations. *Stems*, branched, heavily ascending, semi-woody, rooting at nodes; *Leaves*, alternate, cordate-ovate with milky excretion; *Flowers*, bell-shaped, pink or light violet, ornamental, showy, axillary, borne in bunches almost all along the stem; *Propagation*, by seeds and stem pieces which put forth nodal roots pretty fast.

The weed is aggressive in roadside and farm ditches, waterlogged areas, and shorelines.

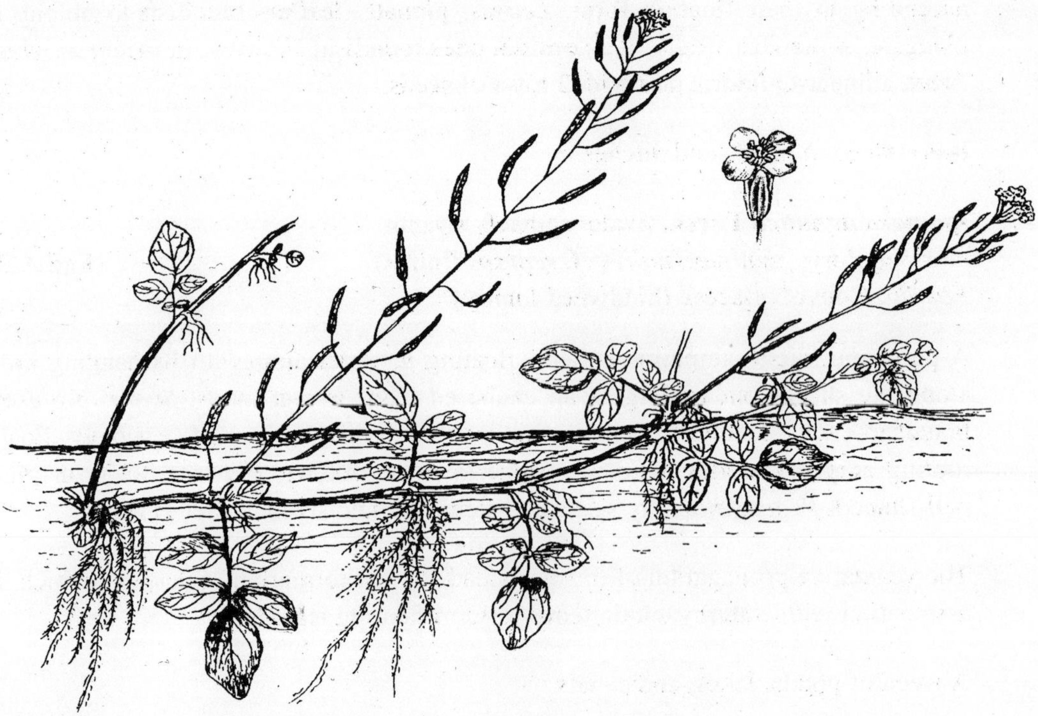

Figure 3.19: *Nasturtium officinale* R. Br. (watercress) (Biotype-2)

Figure 3.20: *Ipomoea aquatica* Forsk. (water spinach) (Biotype-1)

Figure 3.21: *Ipomoea carnea* Jacq. (gramophone flower) (Biotype-4)

19. ***Cabomba aquatica*** Aublet (fenwort, cabomba) **(Fig. 3.22)**
Family: Nymphaeceae (Cabombaceae)

A perennial, dicot, rooted, partly submersed, heterophyllous, up to 2 m high, richly branched herb arising from short rhizomes. *Stems*, with gelatinous coating; *Leaves*, simple, the submersed ones numerous, opposite, finely dissected into thread-like divisions in a hand-fan fashion, 1-3 cm across, the out-of-water ones entire, floating, 6-20 mm long, alternate, oblong to linear-elliptic, peltate, few; *Flowers*, usually solitary, axillary to the floating leaves, white to pinkish, emersed; *Fruit*, a 3-seeded follicle; *Propagation*, by seeds and rhizomes. It is found weedy in lakes, ponds, and quiet streams.

20. ***Hydrilla verticillata*** **(L.F.) Royle** (hydrilla, *jhangi*) **Fig. 3.23 and 3.24**
Family: Hydrocharitaceae (frogbit family)

A monocot, perennial, submersed, rooted herb though its detached stem fragments continue as free-floating plants. *Leaves*, sessile, linear to lanceolate, 0.7-4 cm long, dentate, borne at nodes in whorls of 3-8 throughout the stem; *Flowers*, unisexual, arising singly from axillary spathe, the pistillate ones ascending to the water surface while the staminate ones reaching there by detachment; *Stems*, thin, branched, approaching the water surface even in deep waters; *Roots*, adventitious, filiform; *Propagation*, largely by turions (both aerial and underground) and shoot fragments; *Seeds*, very few.

A very serious weed of lakes, ponds, and canals.

21. ***Nechamandra alternifolia*** **(Roxb.) Thw.**
Family: Hydrocharitaceae (frogbit family)

A submersed, annual or perennial, monocot, dioecious herb. *Stems*, filiform, 60-90 cm long, ascending; *Leaves*, sessile, alternate, linear-lanceolate, 3-8 m long, margins toothed, clasping at the bases; *Flowers*, axillary, in spathes, the male spathes with numerous flowers, submersed, detaching to release their pollens at the water surface, female spathes with solitary, tubular flowers on long peduncles reaching the water surface; *Seeds* numerous, pitted.

A weed of ponds, tanks, and lakes.

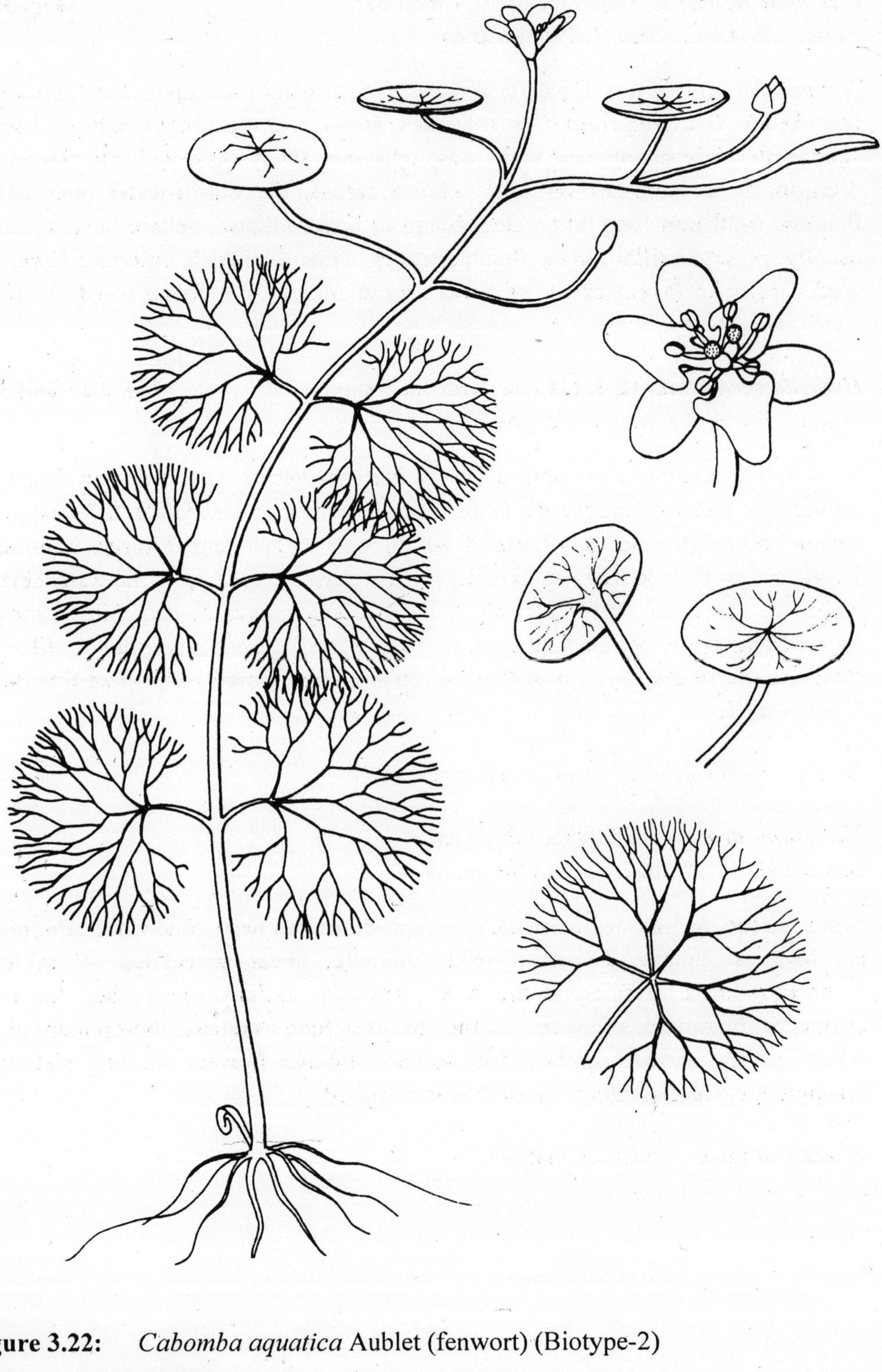

Figure 3.22: *Cabomba aquatica* Aublet (fenwort) (Biotype-2)

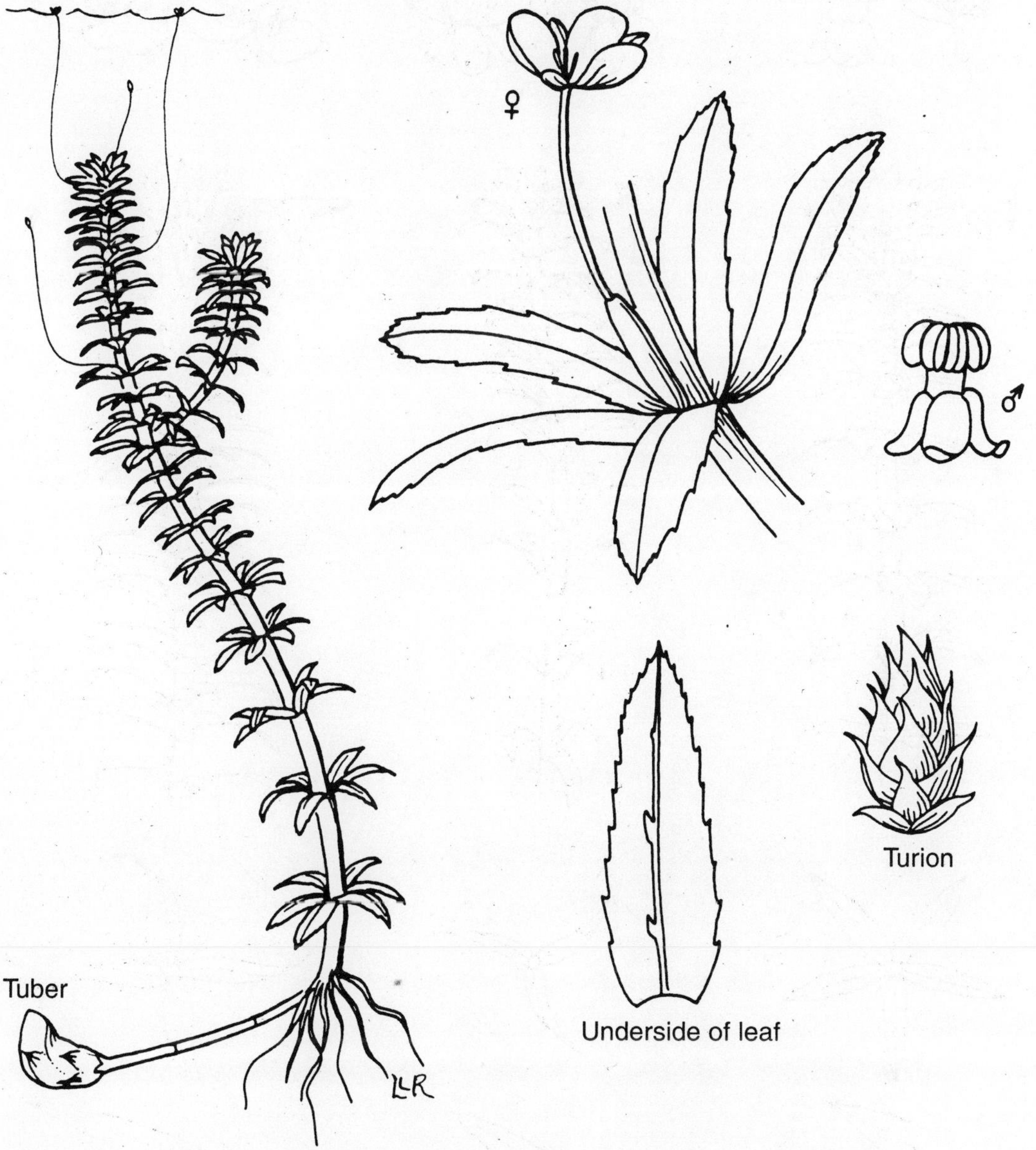

Figure 3.23: *Hydrilla verticillata* (L.F.) Royle (hydrilla) (Biotype-2)

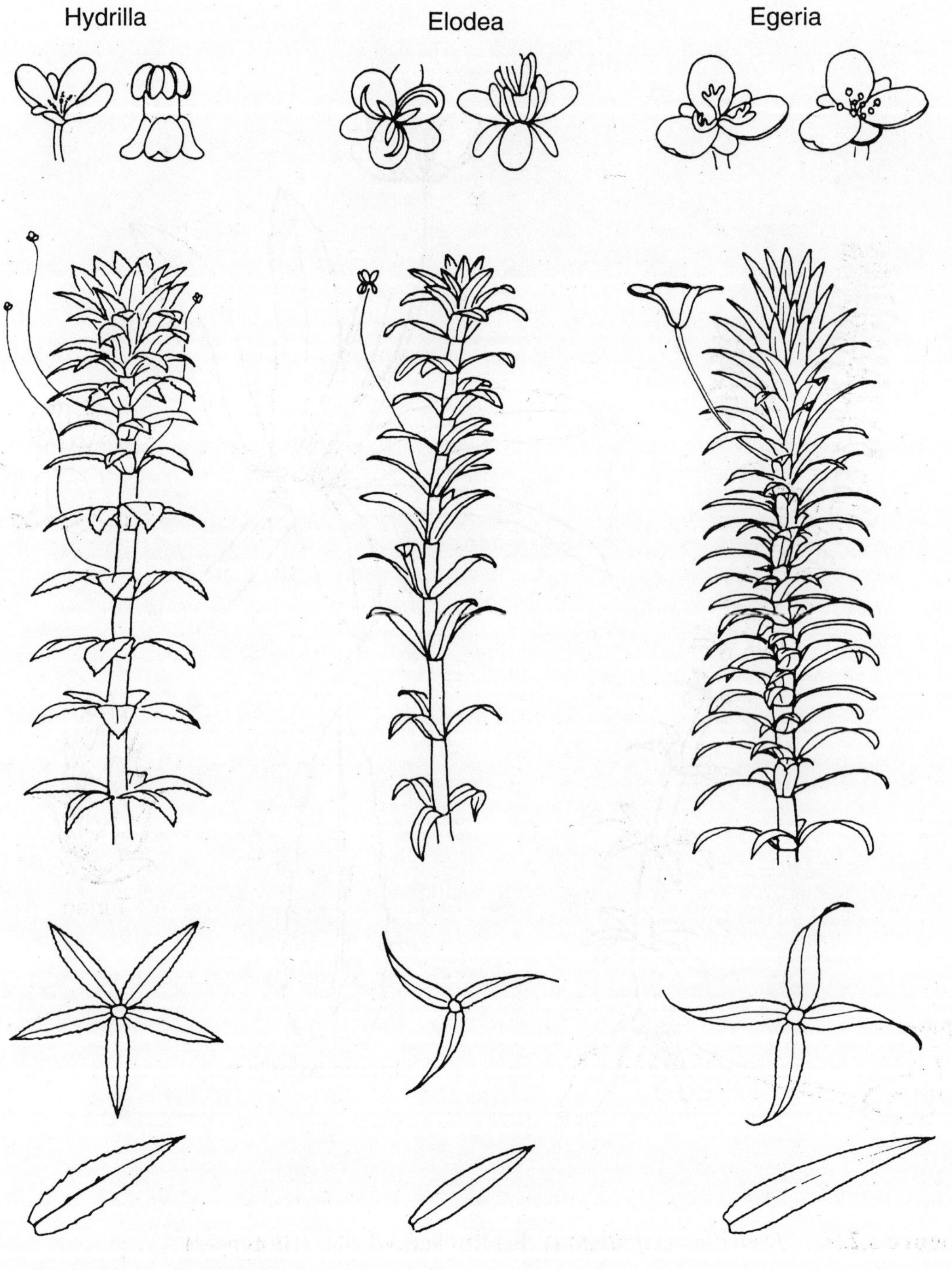

Figure 3.24: Comparative morphology of *Hydrilla, Elodea,* and *Egeria* spp. (Biotype-2)

22.0 *Potamogeton* spp. (pondweeds)
Family: Najadaceae (= Potamogetonaceae) (pondweed family)

Perennial, monocot, submersed, anchored plants, arising from creeping rootstocks. *Leaves*, alternate, heterophyllous, the floating ones of firm texture and the submersed ones thin, membraneous; *Flowers*, largely in spikes, but in some species in heads, rising above the water surface and going beneath it upon maturity; *Propagation*, by rhizomes, tubers, winter buds (turions), and seeds, depending upon the species. Several *Potamogeton* spp. are serious weeds in canals and ponds.

22.1 *P. crispus* Linn. (crisped-leaf pondweed, curyleaf pondweed) **(Fig. 3.25)**

Stem, slender, slackly branched; *Leaves*, all submersed, linear-oblong, blunt or rounded at the end (obtuse), wavy and finely toothed at the margins (crisped), alternate, semi-clasping with dilated bases, 3-nerved, shining (semi-amplexicaul), up to 8 cm long and 0.9 cm wide; *Spike*, emersed, up to 1 cm long, borne on long peduncles, flowers few, sepals clawed.

22.2 *P. nodosus* Poirt. (American pondweed,
longleaf-pondweed, *jalpudedi*) **(Fig. 3.26)**

Stem, thick; *Leaves*, long petioled, heterophyllous, the submersed ones linear, single veined and the floating ones elliptical, multinerved; *Spikes*, axillary, 3.6 cm long, with up to 5 cm long peduncles, densely flowered, flowers sessile, perianth clawed; *Propagation*, by rhizomes, winter buds, and seeds.

22.3 *P. perfoliatus* Linn. (claspingleaf pondweed,
bassweed, red-headgrass)
(= *P. richardsonii* B.R.) **(Fig. 3.27)**

Stem, stout, smooth, cylindrical, slightly branched; *Leaves,* ovate-lanceolate, obtuse, membraneous, shining, clasping at the base (amplexicaul), margins wavy, 5-9 nerved; *Spikes*, densely flowered, up to 2 cm long, emersed, sepals with long claws.

22.4 *P. pectinatus* Linn. (threadleaf pondweed,
bushy pondweed, sago pondweed) **(Fig. 3.28)**

Stems, filiform, copiously branched, *Leaves*, all submersed, thread-like, acute, up to 15 cm long, spread under water in a hand fan fashion to give plant a bushy appearance; *Spikes*, axillary, emersed with sessile flowers placed like a string of beads; *Propagation*, by rhizomes, tubers, and seeds.

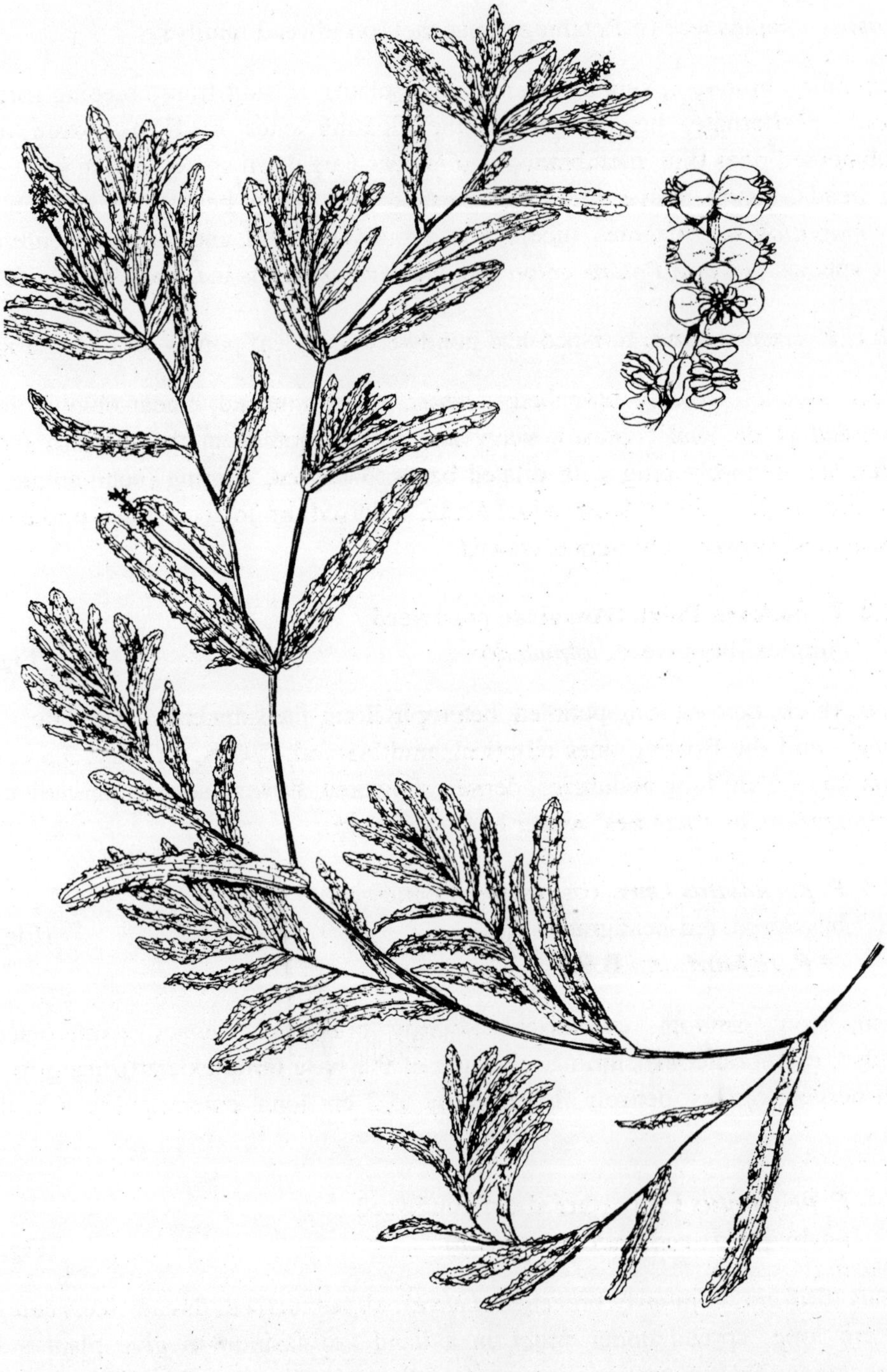

Figure 3.25: *Potamogeton crispus* Linn. (crisped-leaf pondweed) (Biotype-2)

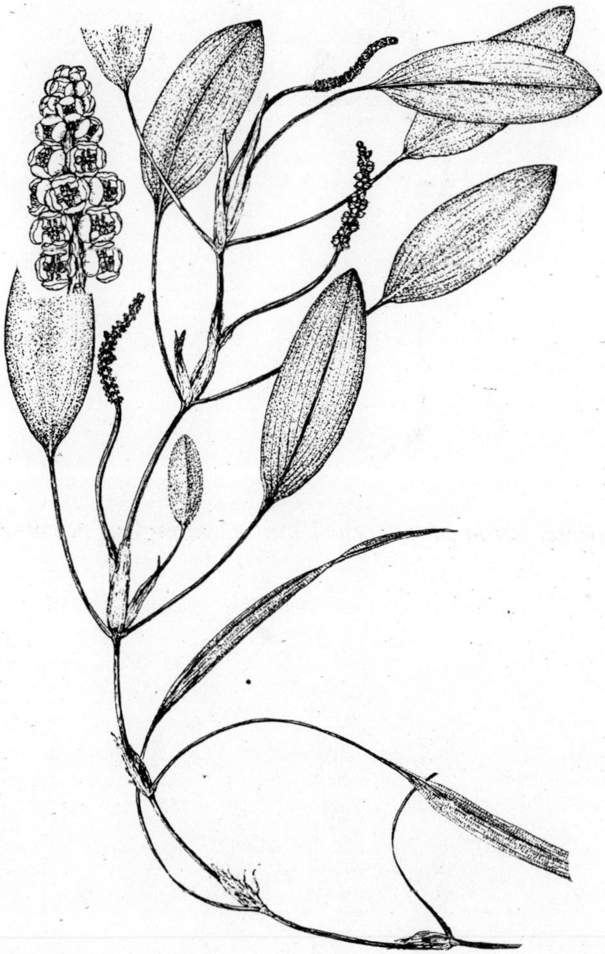

Figure 3.26: *Potamogeton nodosus* Poirt. (longleaf pondweed) (Biotype-2)

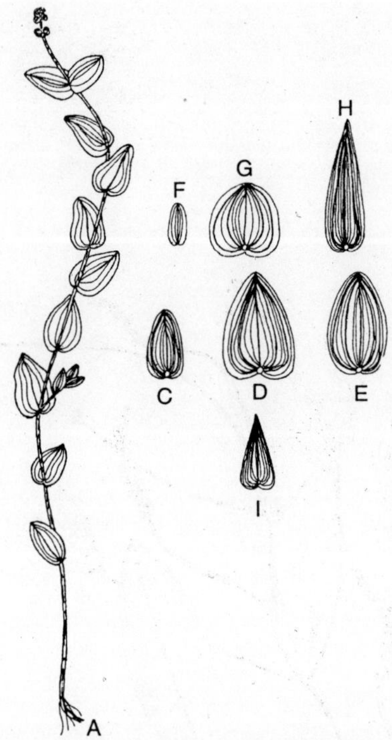

Figure 3.27: *Potamogeton perfoliatus* Linn. (claspingleaf pondweed) (Biotype-2)

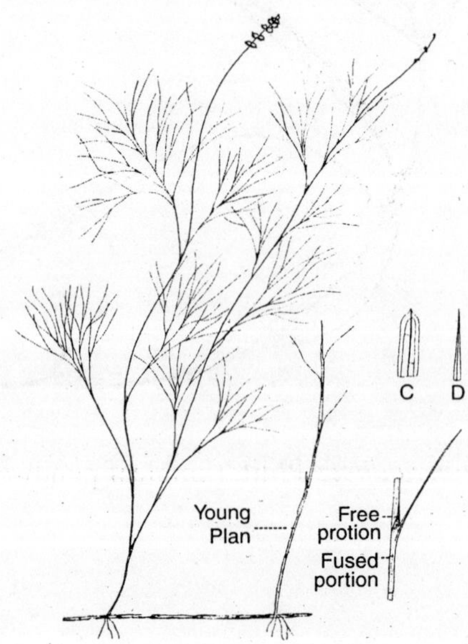

Figure 3.28: *Potamogeton pectinatus* Linn. (threadleaf pondweed) (Biotype-2)

23.0 *Najas* spp. (naiads)
Family: Najadaceae (pondweed family)

Annual or perennial, submersed, anchored, monocot plants. *Stems*, slender, much branched; *Leaves*, opposite, ribbon-like, enlarged at the base; *Flowers*, monoecious, axillary, in singles; *Fruit*, single seeded; *Seed*, tapering at either end, enclosed in a papery envelope (*cf. Zannichellia* spp.); *Habitat*, same as of *Potamogeton* spp.

23.1 *N. minor* All. (brittle naiad, ***jhanji***) **(Fig. 3.29)**

A delicate, annual aquatic submersed plant; *Stems*, 20-30 cm long, very thin, fork-shaped, *Leaves*, narrow, linear, 1-2 cm long, sickle-shaped with 6-10 pairs of teeth, often recurving with fine teeth above and smooth below, widening of leaf bases lobe-like, the lobe finely toothed and fan-shaped; *Fruit*, a 2-3.5 mm long, cylindrical achene; *Seed*, with broad pits in vertical rows.

23.2 *N. graminea* Delile

Stem, mucricate with 36-40 teeth on the margins, rooting at lower nodes; *Fruit*, narrow-oblong or ellipsoid-oblong, tapering at apex; *Seed*, small, areolated with 25-30 lines.

24. *Zannichellia palustris* L. (horned pondweed) **(Fig 3.30)**
Family: Najadaceae (Zannichelliaceae)

Stems, numerous, thread-like, from creeping rootstocks; *Leaves*, thread-like, opposite; *Flowers*, axillary, with short stalks; *Fruits* in branches of 2-5, 2-3.0 mm long, flattish, dentate on one side; *Seeds*, enclosed in a fruit coat which is narrowed to a stipe at one end and a slender style at the other, and toothed down on one side; *Habitat*, same as of *Najas*. It can withstand brackish waters.

25. *Ceratophyllum demersum* L. **(Fig. 3.31)**
(hornwort, common coontail)
Family: Ceratophyllaceae (hornwort family)

A perennial, submersed plant, usually without roots, anchored to hydrosoil by shoots in shallow waters but sometimes floating on the deep water surfaces; *Stems*, elongate, freely branched, forming large masses; *Leaves*, 1-3 cm long, borne in whorls of 5-12 each, divided palmately 2-3 times in a fork-like fashion, leaflets 0.5 mm wide, conspicuously serrate on one side, often more crowded towards the apical ends of branches; *Flowers*, unisexual, minute, solitary, sessile, axillary, each subtended by a calyx-like involucre of 8-14 bracts; *Fruit*, an achene, ellipsoid, 4-6 mm in length, with 2 basal spines; *Propagation*, by seeds and stem fragmentation.

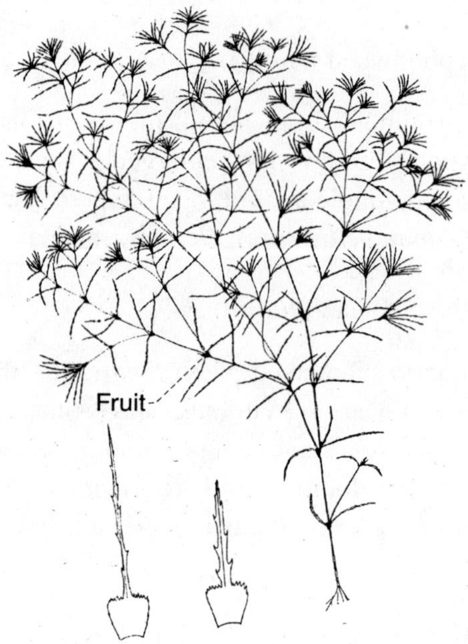

Figure 3.29: *Najas minor* All. (brittle naiad) (Biotype-2)

Figure 3.30: *Zannichellia palustris* L. (horned pondweed) (Biotype-2)

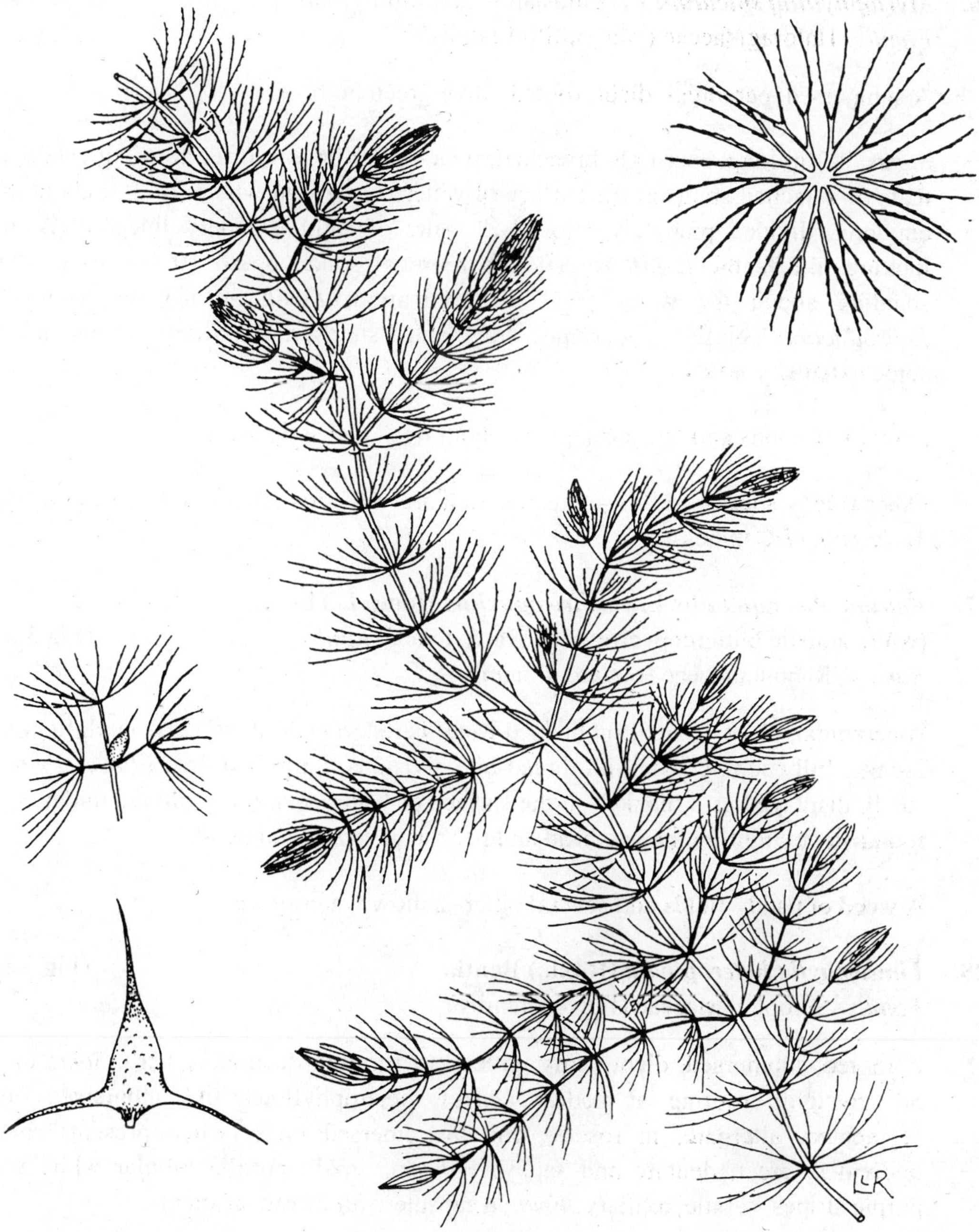

Figure 3.31: *Ceratophyllum demersum* L. (common coontail) (Biotype-2)

A common weed of quiet waters and ponds.

26. ***Myriophyllum spicatum* L.** (Eurasian watermilfoil, *jhanji*) **(Fig. 3.32)**
Family: Haloragidaceae (watermilfoil family)

A submersed, perennial, dicot, rooted, olive-green herb.

Stems, 2-3 m long, sparingly branched, wine-red or brown, often forming extensive mats of detached stems at the surface of water. *Leaves*, in whorls of 3-4, about 3-4 cm long, divided pinnately into 14-21 pairs of rigid, slendery linear divisions, looking like feathers; *Inflorescence*, a terminal spike, up to 10 cm long, often standing above the water level till pollination; *Fruit*, an achene, 4-jointed; *Reproduction,* chiefly by rhizomes, fragmented stems, and axillary turions and, to some extent, by seeds.

A weed of ponds and lakes, adapted to both fresh and saline waters.

Other weedy species in this genus reported are: *M. aquaticum* (Vellose) Verdc. and *M. brasiliens* Cambess.

27. ***Ranunculus aquatalis* L. *var. tricophyllus* Hook. f. Thom.**
(white aquatic buttercup, crowfoot, water-fennel, *tolub*) **(Fig 3.33)**
Family: Ranunculaceae (buttercup family)

A perennial, dicot, rooted, submersed water herb. *Stem*, hollow, horizontal, floating; *Leaves*, fully submersed, with thread-like divisions, limp, petioled; *Flowers*, white, small, displayed as a blanket on the surface of water; *Fruit*, a wrinkled nutlet, with roundish or slightly elongate head and persistent sepals and petals.

A weed of paddy fields and several other shallow water places.

28. ***Limnophylla heterophylla* (Roxb.) Benth.** **(Fig 3.34)**
Family: Scrophulariaceae (figwort family)

A rooted, submersed, dicot, leafy, green tuft forming decorative herb. *Stem*, up to 50 cm long, rooting at nodes; *Leaves*, heterophyllous, the submersed ones segmented, alternate, in rosette, and the emersed ones (when present) small, undivided, erect, dentate and sappy; *Flowers,* small, corolla tubular white with purple stripes, sessile, axillary; *Fruit*, a capsule with 2 compartments.

A weed of tanks, ponds, and shallow streams.

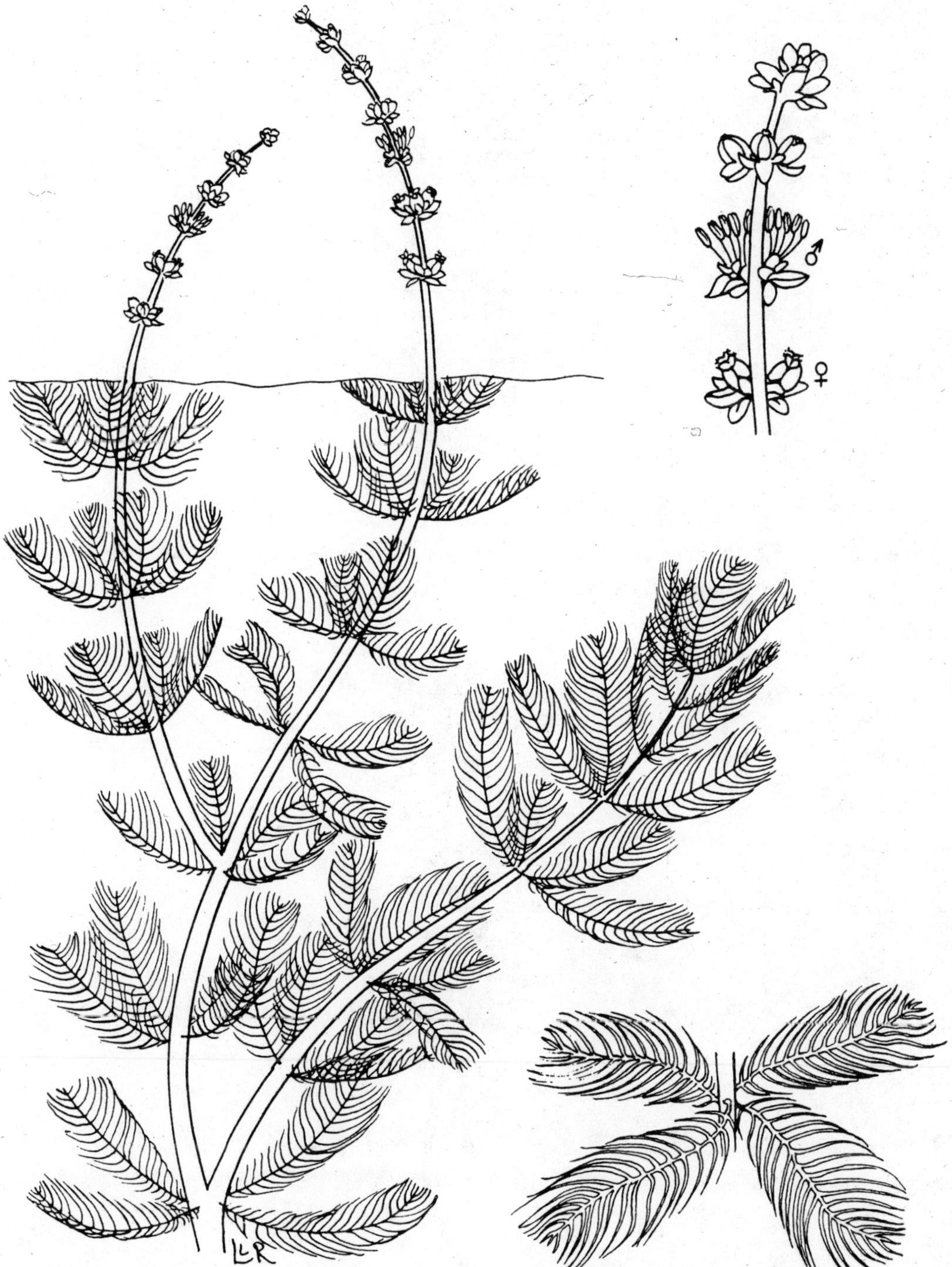

Figure 3.32: *Myriophyllum spicatum* L. (Eurasian watermilfoil) (Biotype-2)

Figure 3.33: *Ranunculus aquatalis* L. *var. tricophyllus* Hook. F. Thom. (white aquatic buttercup) (Biotype-2)

Figure 3.34: *Limnophylla heterophylla* (Roxb.) Benth. (Biotype-2)

29. ***Utricularia vulgaris*** **L.** (bladderwort) **(Fig 3.35)**
Family: Lentibulariaceae (bladderwort family)

A perennial, rootless herb floating beneath the surface of water. *Stem*, thin, 30-100 cm long, with green and coarse branches which divide six or more times to give it a leaf-like appearance, branches carry axillary bladders (insect traps); *Flowers*, yellow, 2-lipped, borne on 10-30 cm tall scape, elevated above the water surface; *Fruit*, a several-seeded pod; *Propagation*, by 1-2 cm long winter buds (turions).

A weed of still waters in ponds and ditches.

Other important species of *Utricularia* found weedy in India include *U.stellaris* L. f., *U. gibba* L., *U. minor*, *U. gemniscapa* Benj., *U. flexuosa* Vahl Erium., and *U. exoleta* R. Br.

30. ***Ottelia alismoides*** (L.) Pers. **(Fig 3.36)**
Family: Hydrocharitaceae (frogbit family)

It is an annual, submersed, monocot herb with ornamental features. *Leaves*, heterophyllous, the initial ones ribbon-like and the subsequent ones round to lanceolate, bulging and wavy, borne on 10-20 cm long petioles; *Blossom*, solitary, appearing through hollows in the bipartite bracts, scapigerous, white and floating on the water surface. *Fruit*, a capsule; *Propagation*, by seeds.

This plant is weedy largely in lowland paddy fields in southern and eastern India, besides being a shoreline weed of several ponds.

31. ***Sparganium erectum*** L. (burreed) **(Fig 3.37)**
Family: Sparganiaceae (burreed family)

A monocot, perennial herb of marshlands and swamps.

Leaves, erect, all basal from rhizomes, keel-shaped, long, 3-15 mm wide; *Inflorescence*, axillary or supra-axillary, roundish with clusters of sessile flowers, heads arranged at intervals on shoots, lower heads pistillate and the upper ones staminate; *Propagation*, by nutlets (partly surrounded by scale-like sepals) and creeping rootstocks.

It is a weed of lowland paddy fields, shores, and ponds.

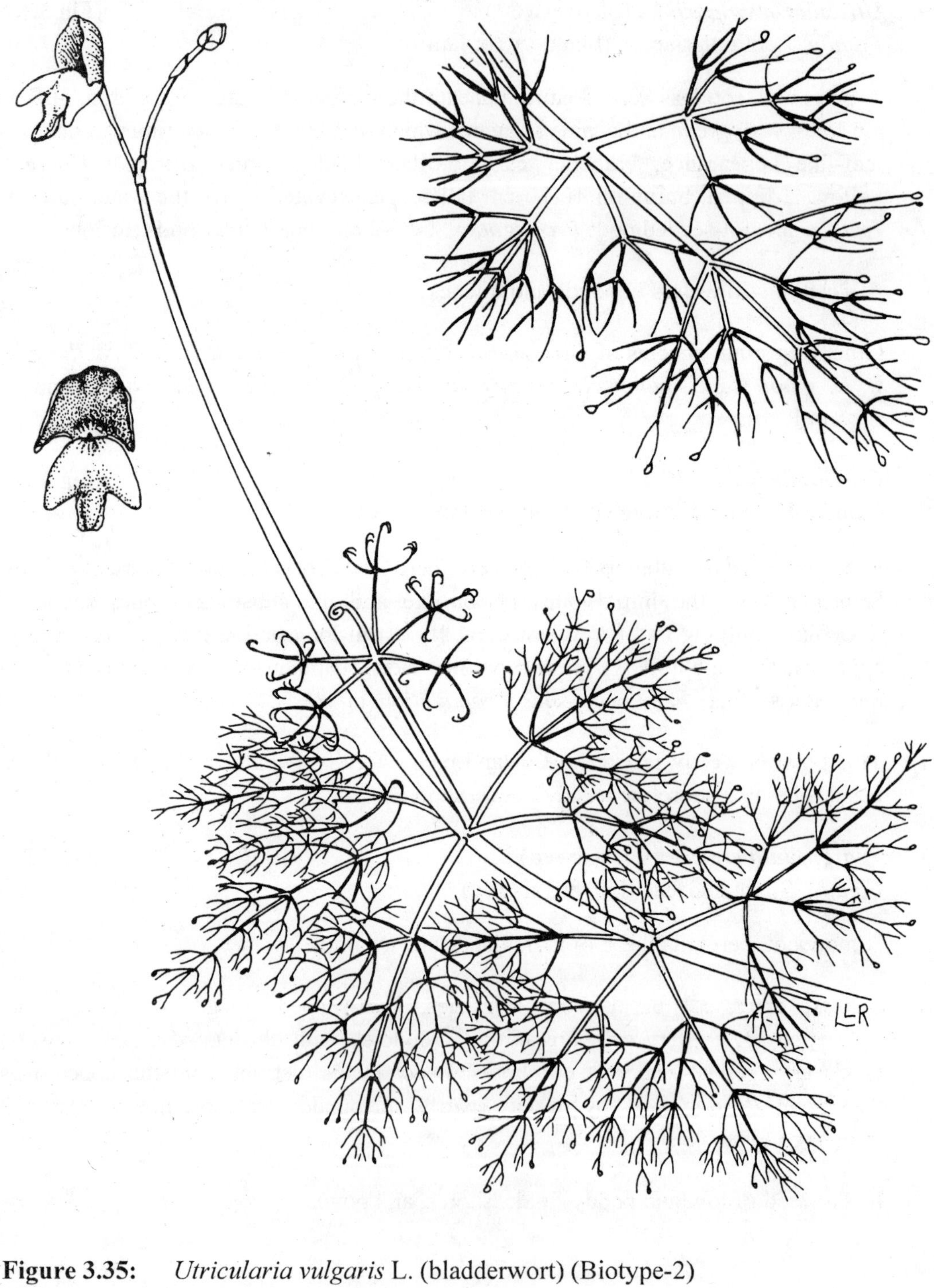

Figure 3.35: *Utricularia vulgaris* L. (bladderwort) (Biotype-2)

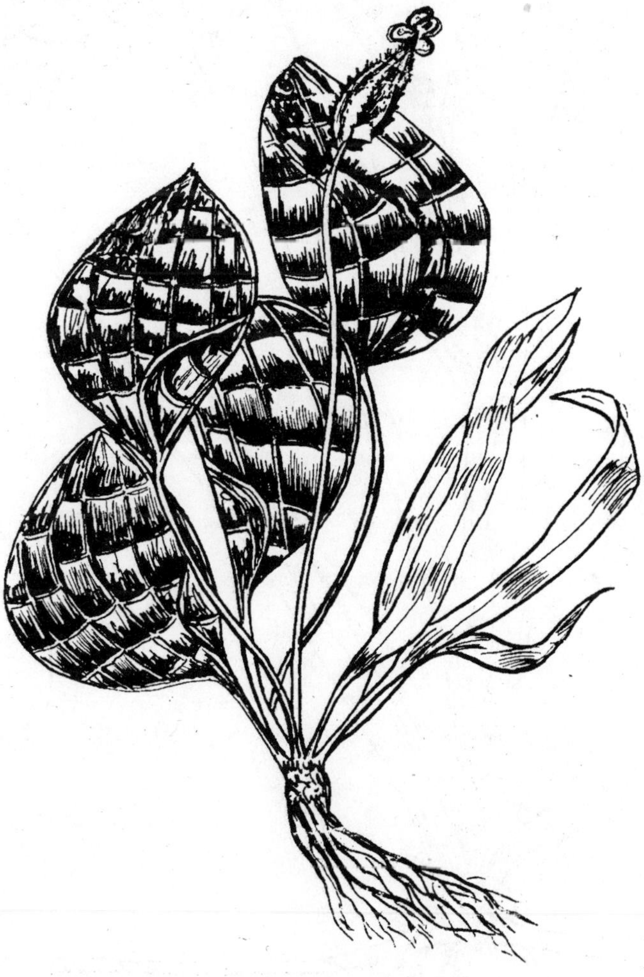

Figure 3.36: *Ottelia alismoides* (L.) Pers. (Biotype-2)

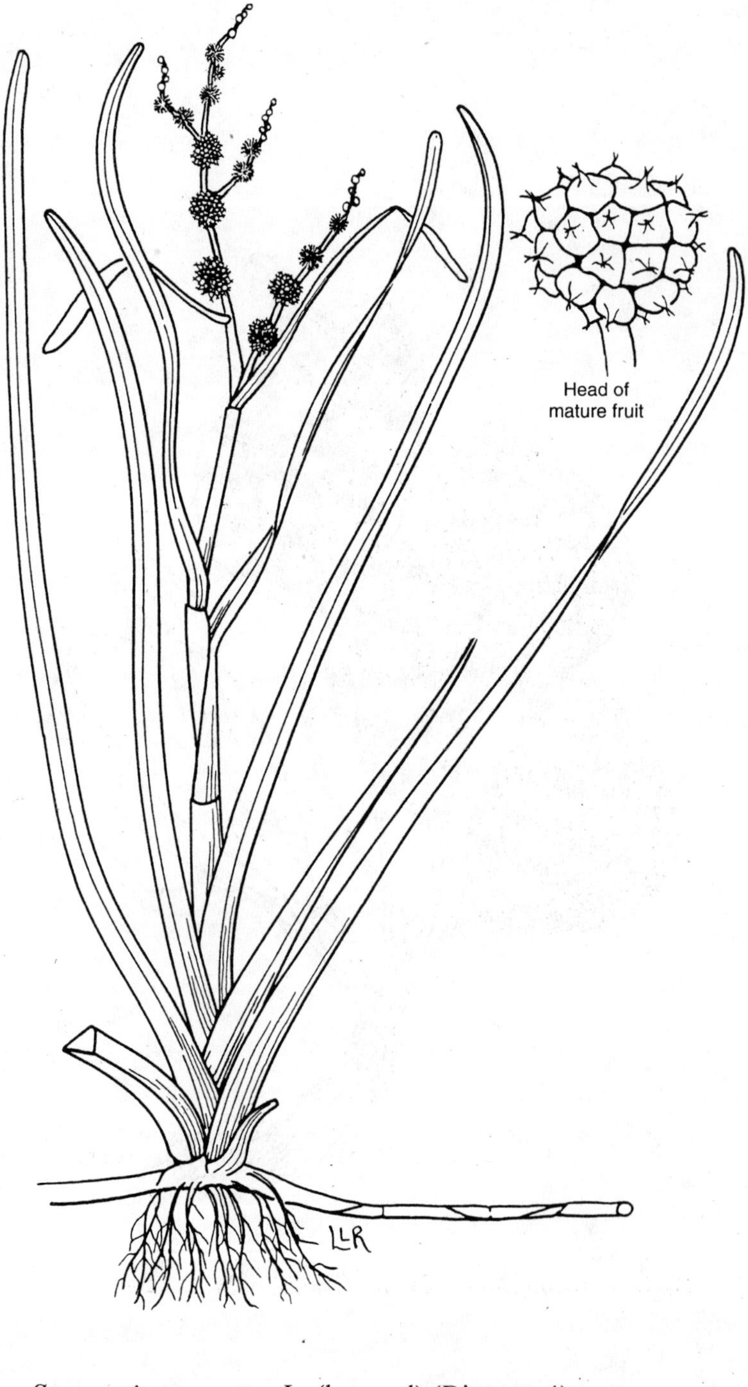

Figure 3.37: *Sparganium erectum* L. (burreed) (Biotype-4)

32. ***Vallisneria spiralis*** **L.** (tapeweed, eelgrass, *patasavala* wildcelery, *savala*)**(Fig 3.38) (= *V. natans* (Lour.) H. Hara).**
Family: Hydrocharitaceae (frogbit family)

Perennial, dioecious, monocot, completely submersed leafy plant. *Stems*, slender, creeping inside mud as rootstocks, with fibrous roots at each node; *Leaves*, in rosettes of 5-20, light-green, narrow (up to 1.5 cm wide), linear (up to 70 cm long), ribbon-like, with 5 longitudinal veins ending parallel to the leaf tip (*cf. Sagittaria*), tips with tiny dentations, bases reddish brown or reddish–blue; *Flowers*, male plants with numerous, tiny, 2-stamens flowers placed at the plant bases, wrapped in a spathe with floating arrangements, female plants with solitary, pistillate flowers borne in tubular pinkish-white spathes on long, coily peduncles reaching the water surface. Male flowers upon detachment rise to the surface of water to fertilise the female flowers. The peduncles retract after pollination in a coily fashion; *Fruit*, a cylindrical, 15-12 mm long capsule.

This weed infests chiefly the slow moving and still waterbodies like lakes, ponds, and shores of canals and rivers.

33. ***Butomus umbellatus*** L. (flowering rush) **(Fig 3.39)**
Family: Butomaceae (flowering rush family)

A perennial, monocot, emersed plant of shallow waterbodies. *Leaves*, ribbon-like, arising from stout rootstocks, erect in shallow waters and limping when fully submersed, up to 45 cm long; *Inflorescence*, an umble, terminating on erect, about 1 m tall scape, pink, very showy; *Fruit*, a follicle with long beak; *Propagation*, by seeds and rootstocks, sterile biotypes on record.

34. ***Sagittaria sagittifolia*** L. (arrowhead, duck-potato) **(Fig 3.40)**
Family: Alismaceae (water-plantain family)

A perennial, monocot, anchored, basically emersed herb. *Stems*, underground (rhizomes), ending in edible tubers; *Leaves*, growing from rhizomes in rosettes, leaf blades 10-30 cm in length, arrowhead-shaped with long petioles, basal lobes pointing backwards (sagittate), commonly extending above the water surface though can survive submergence. *Inflorescence*, a scapigerous spike towering above the leaves, comprising usually three whorls of 3-white flowers and 3 small bracts each arranged at short intervals, staminate flowers above and the pistillate below, female blossoms with numerous clustered ovaries in spherical formation; *Roots*, fibrous; *Fruits*, beaked nutlets; *Propagation*, by seeds and rhizomes.

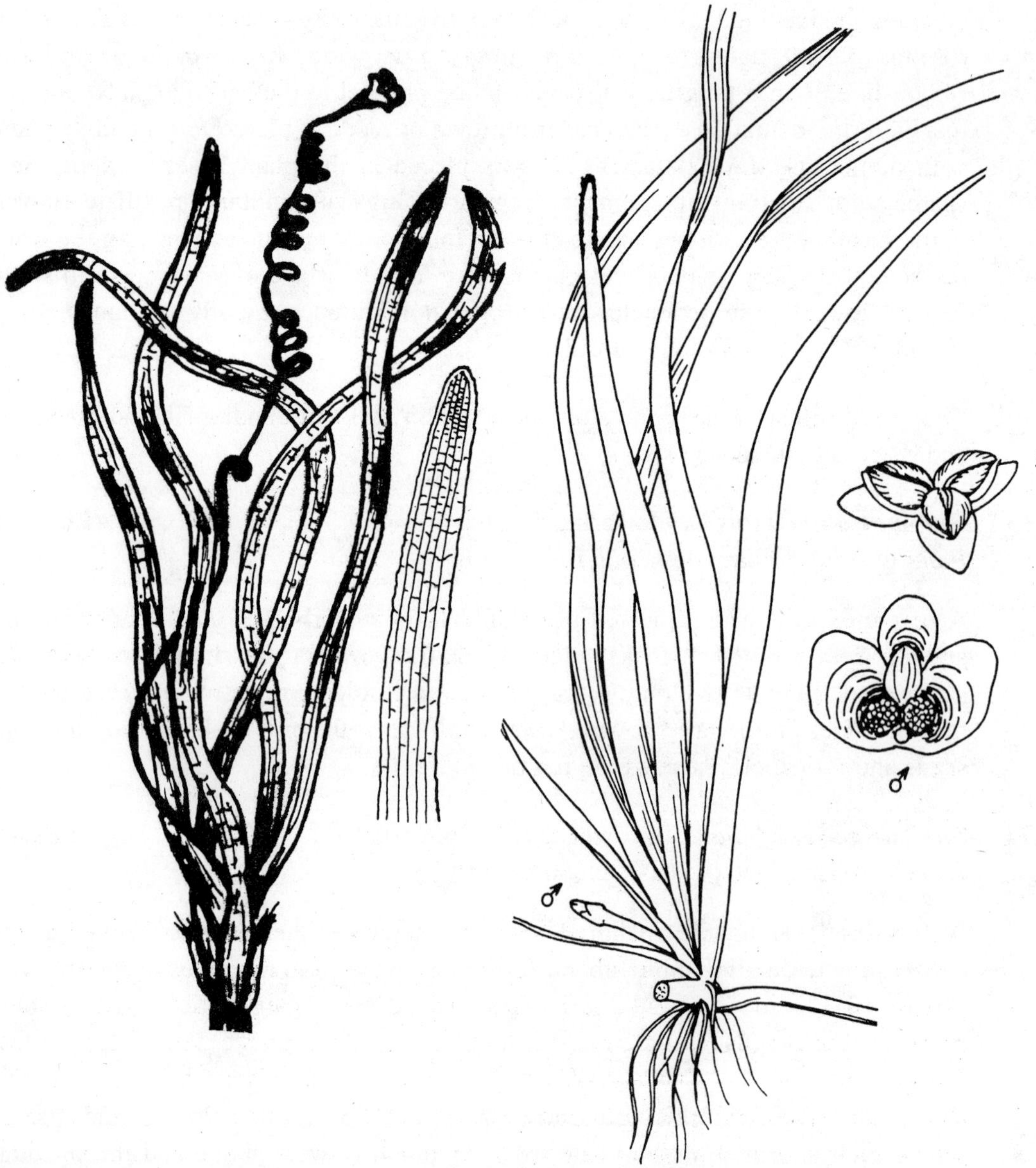

Figure 3.38: *Vallisneria spiralis* L. (tapeweed) (Biotype-2)

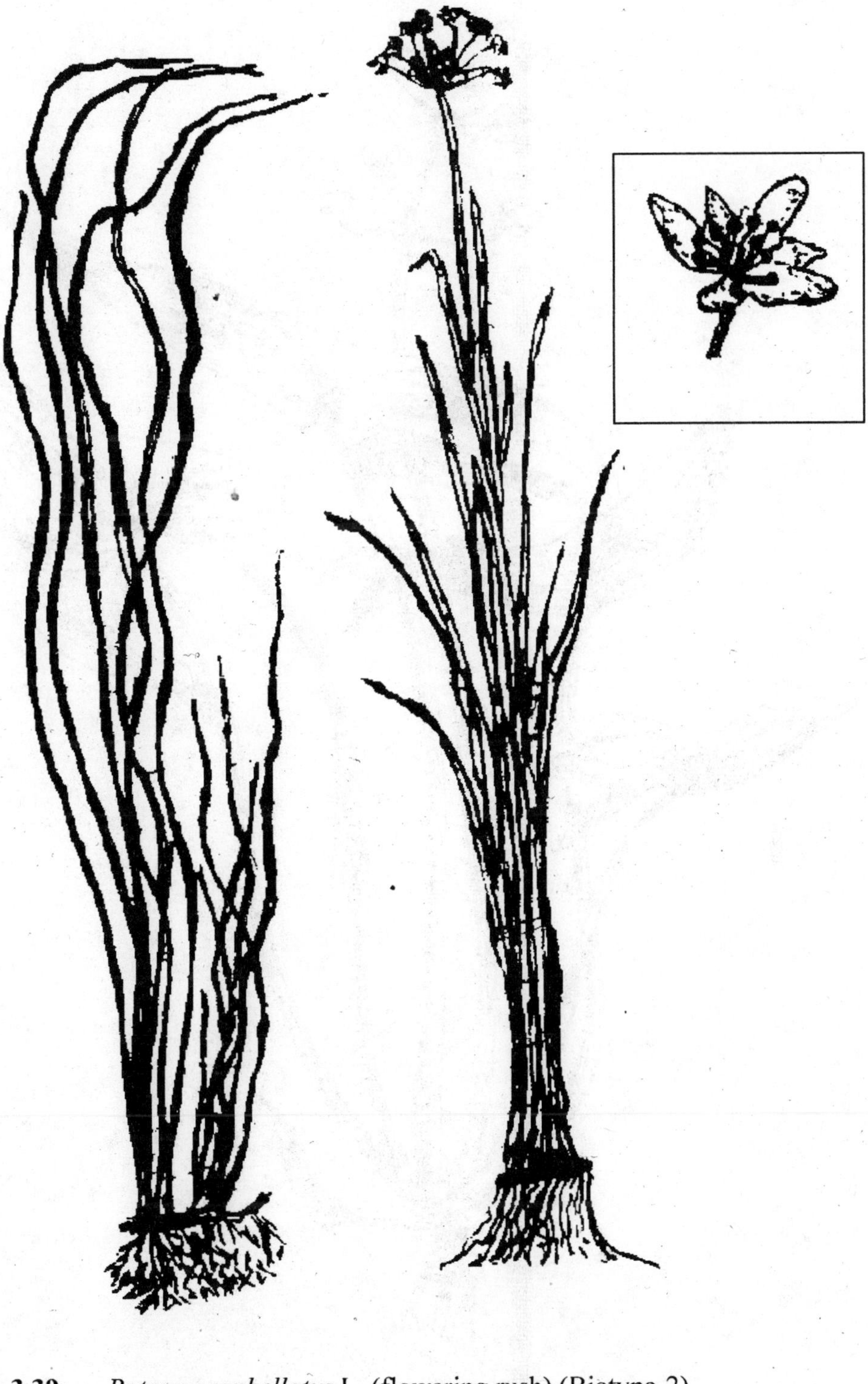

Figure 3.39: *Butomus umbellatus* L. (flowering rush) (Biotype-2)

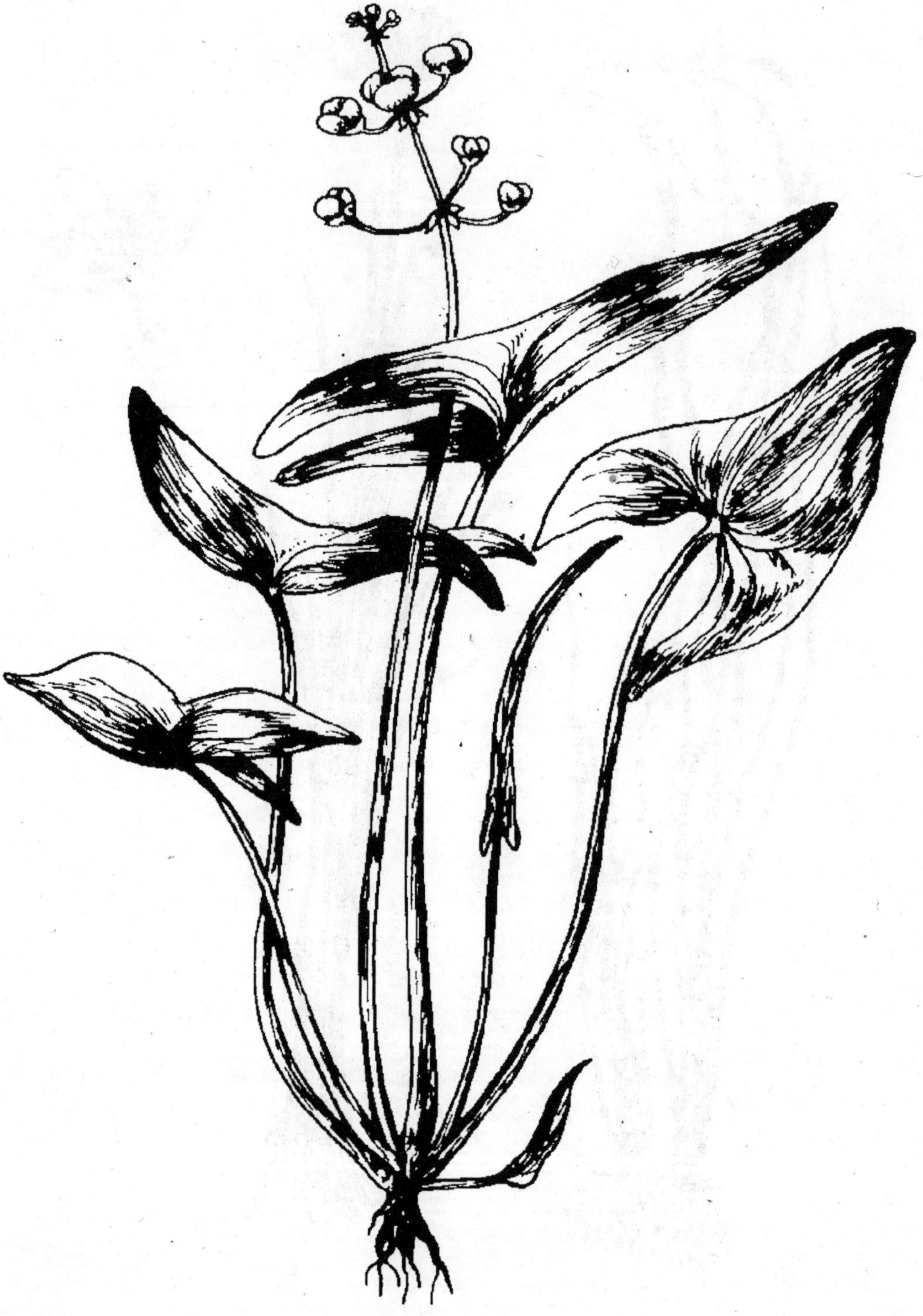

Figure 3.40: *Sagittaria sagittifolia* L. (arrowhead) (Biotype-3)

A weed of chiefly drainage ditches, lowland paddies, shores, tanks, and like shallow water situations.

35. ***Colocasia antiquorum* Schott.** *(kachalu)* **(Fig 3.41)**
(= *C. esculenta* (L.) Schott.)
Family: Araceae

A perennial, monocot, robust, erect, aquatic ally plant. *Stem,* thick, up to 1 m high often spreading by stolons; *Leaves*, very large, smooth, oblique, sagittate, peltate with prominent mid-vein forked at the base, borne on up to 2 m long petioles sheathing the stems at the bases; *Inflorescence*, a spadix with a long, ovate-lanceolate spathe, male flowers above and female flowers below, some flowers sterile; *Rhizomes*, tuberous, starch-filled, often used as vegetable.

An aggressive weed of humid areas choking badly the open drains and channels and blocking the vast damp lands.

36. ***Alisma plantago-aquatica* L.** (waterplantain) **(Fig 3.42)**
Family: Alismaceae (waterplantain family)

A stout, perennial, monocot bog plant, 10-20 cm high, from tuberous rootstocks. *Leaves*, heterophyllous, in submersed state limp, ribbon like, about 1 m long, but in terrestrial form heart-shaped, firm, with long petioles; *Inflorescence*, up to 1 m high, scapigerous, with small, white or rose coloured blossoms arranged in threes.

A weedy plant in drainage ditches where, alongwith *Sagittaria, Sparganium*, and *Typha* spp., its emerged form impedes the water flow.*

37. ***Aponogeton crispus* Thunb.** **(Fig. 3.43)**
(= *A. undulatus* Roxb.)
Family: Aponogetonaceae

It is a perennial, monocot, submersed plant. *Leaves*, borne on distinct 30-35 cm long petioles from tuberous, thick rootstocks, blades up to 30 cm long and 3-5 cm wide, gently crispate; *Inflorescence*, an emersed spike, scapigerous, with creamy flowers.

---

* ***Note:*** *Spadix: A fleshy spike of flowers.*

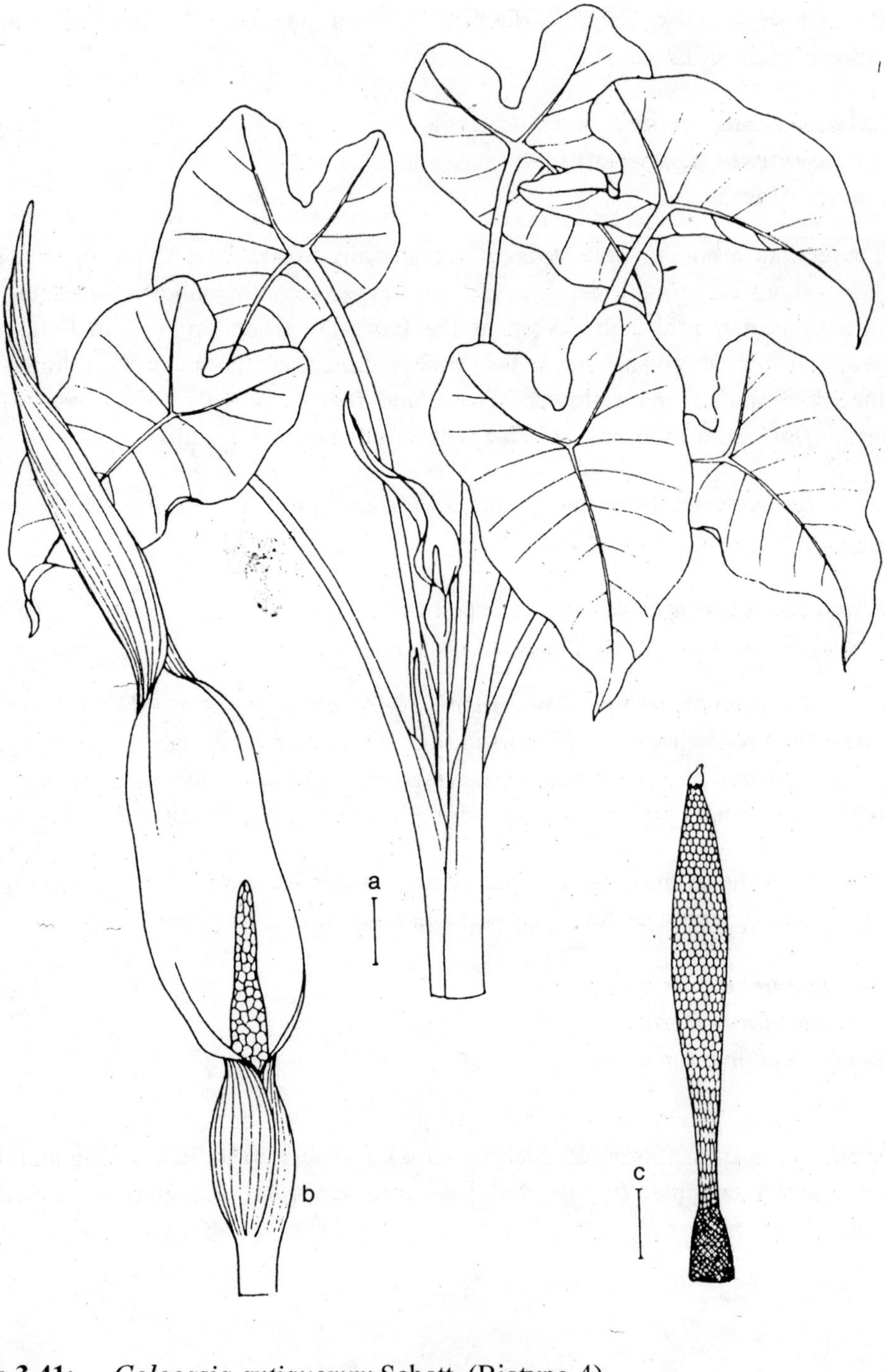

Figure 3.41: *Colocasia antiquorum* Schott. (Biotype-4)

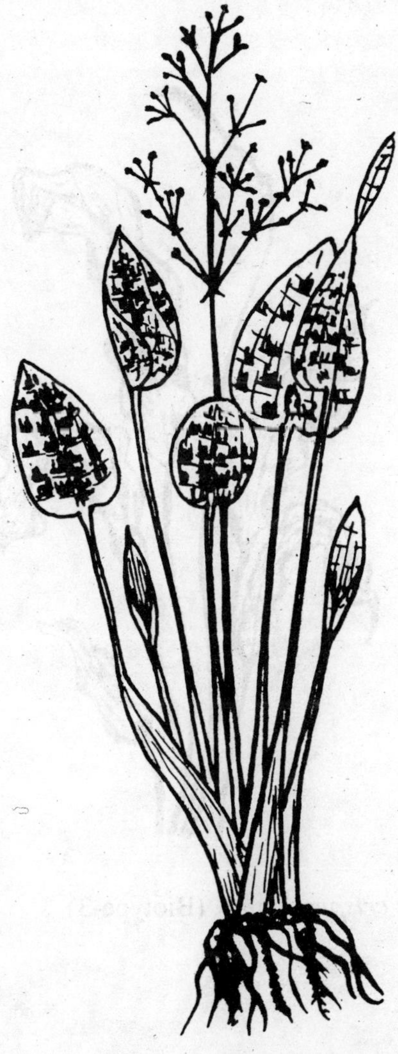

Figure 3.42: *Alisma plantago-aquatica* L. (waterplantain) (Biotype-3)

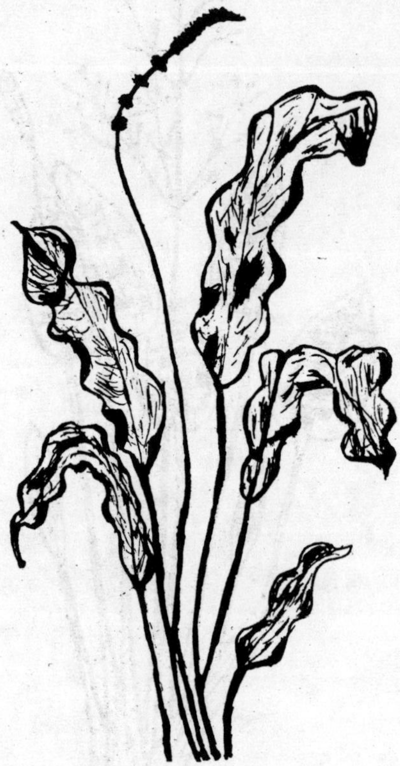

Figure 3.43: *Aponogeton crispus* Thunb. (Biotype-3)

This specie crosses frequently with *A. natans* (L.) E. & K. which possesses white or bluish flowers. In nature the pure species of *Aponogeton* are difficult to find and its hybrids are sterile, propagating only by rootstocks.

A weed of ponds, tanks, and shallow lakes.

38. *Monochoria vaginalis* (Burm. f.) Presl ex. Kunth
(meerthomari, nanka, panari) **(Fig. 3.44)**
Family: Pontederiaceae (pickerelweed family)

A perennial, clump forming, apparently stemless, emersed, aquatic herb with a suberect, short rhizome. *Leaves*, arising from rhizomes with 10-15 cm long petioles, blades of erect leaves broadly lanceolate but the floating ones sometimes linear; *Flowers*, blue, sprinkled with red.

A serious weed of lowland rice fields and marshlands in Eastern India in particular.

39. *M. hastata* (L.) Solms **(Fig. 3.45)**

It is similar in growth behaviour and habitat to *M. vaginalis* but it is easily differentiated by its hastate leaves, long petioles, creeping, branched, well developed spongy rhizomes, and violet-blue flowers with red dots.

40. *Juncus effusus* L. (softrush, matrush) **(Fig. 3.46)**
Family: Juncaceae (rush family)

An erect, about 1 m tall plant with large clumps of several hundred cylindric, pale-green stems arising from rhizomes, often confused with a grass or a sedge. There are sheaths at the base of the stems which are not terminated by leaf blades. *Inflorescence*, terminal but looking lateral because of the presence of a long, erect, involucral leaf appearing as continuation of the stem beyond the inflorescence; *Flowers,* lily-like with three scaly sepals and three petals without scales (*cf. Scirpus*); *Propagation*, by rhizomes and seeds.

A weed of paddy fields, shorelines, and ditches. Some true water species of *Juncus* include *J. repens*, *J. subtilis*, and *J. pelocarpus*.

41.0 Grasses
Family: Poaceae (= Gramineae)

Grasses are characterised by their cylindrical, hollow, unbranched stems, 2-rank leaves which are longer than wide, open sheaths, presence of a ligule of variable shapes and sizes at the junction of leaf blade and sheath, and mealy grains (fruits).

Figure 3.44: *Monochoria vaginalis* (Burm. f.) Presl ex. Kunth (Biotype-3)

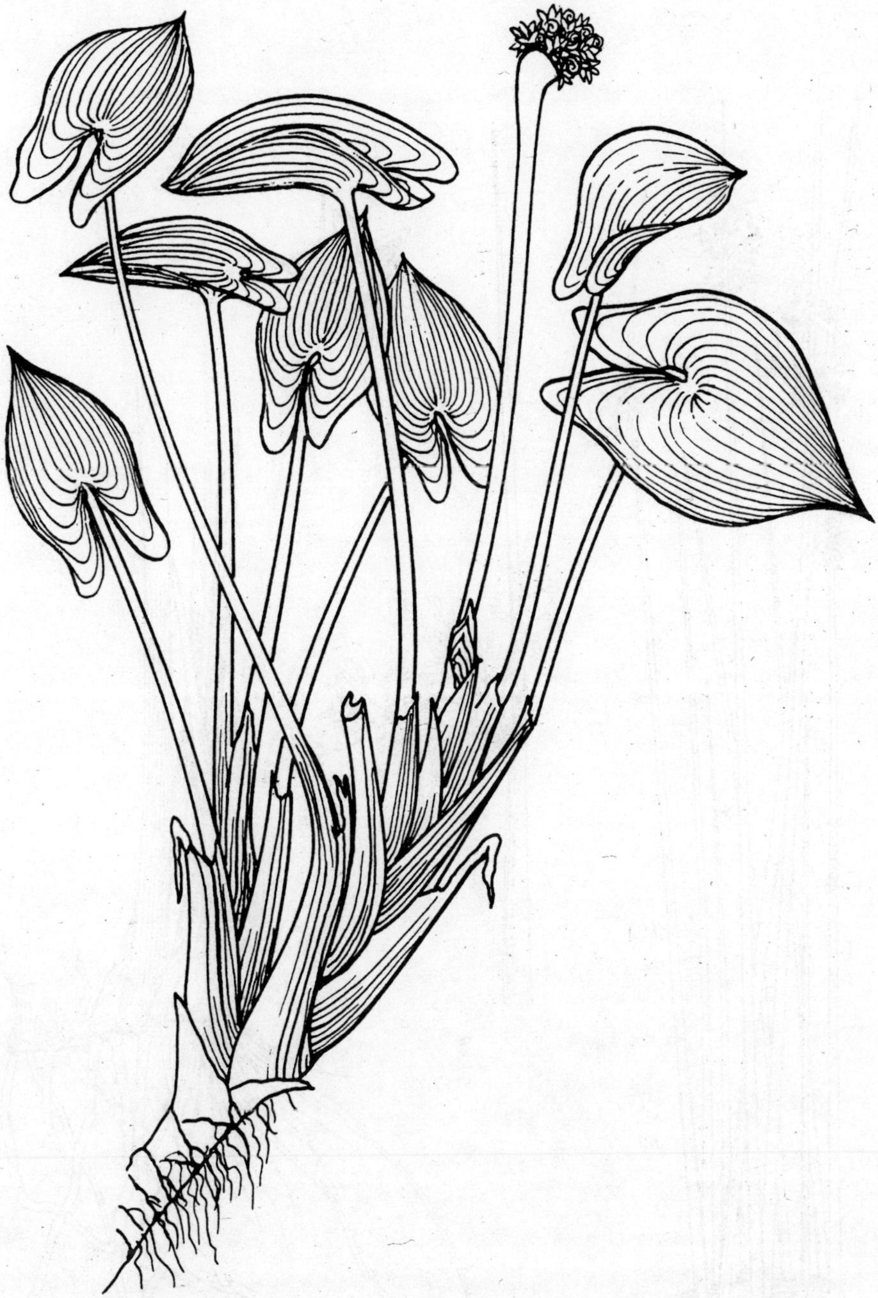

Figure 3.45: *Monochoria hastata* (L.) Solms (Biotype-3)

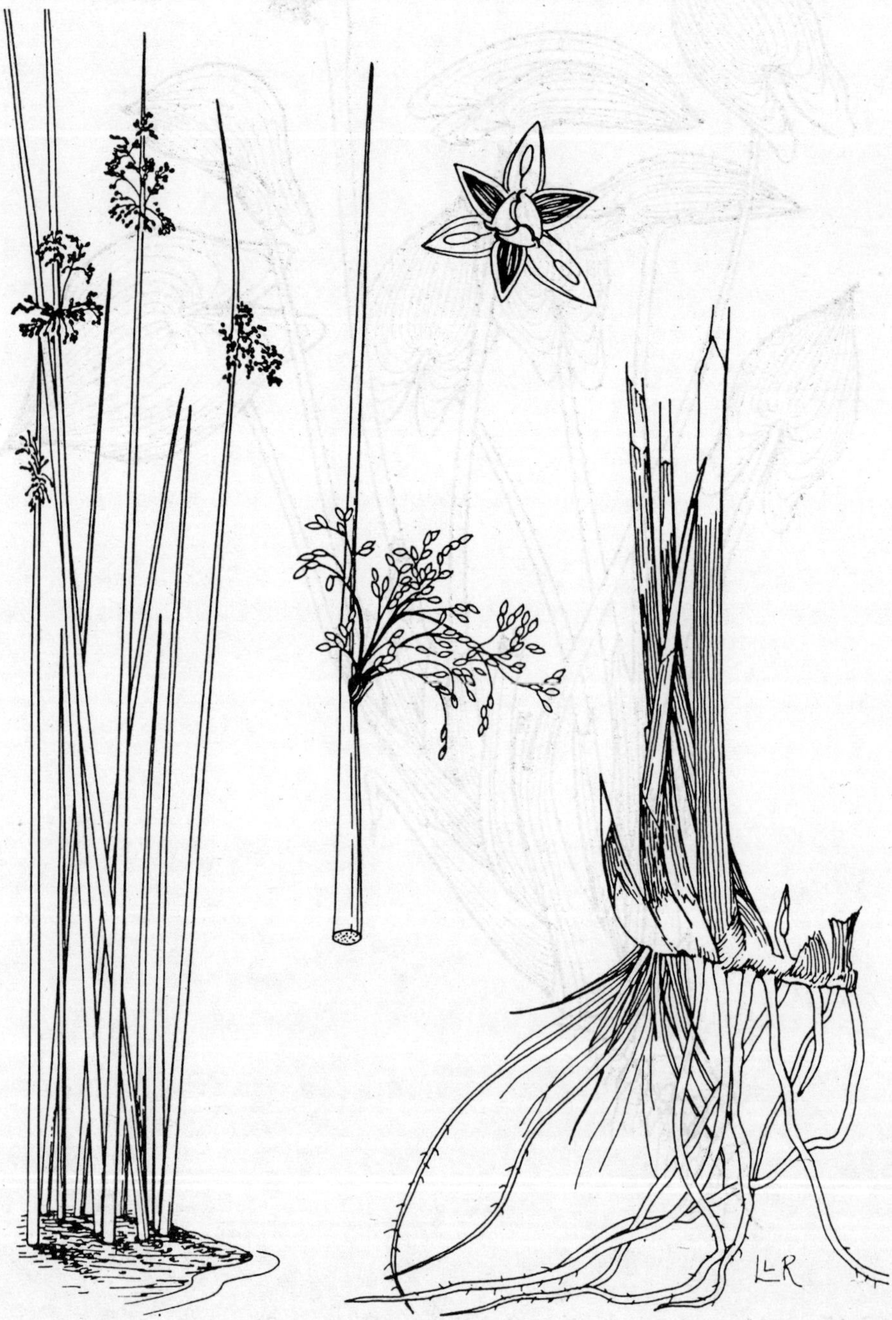

Figure 3.46: *Juncus effusus* L. (softrush) (Biotype-4)

41.1 *Phragmites karka* (Retz.) Trin. ex. Steud.

(common reed, marshgrass, *narkul, dila, madrukahati*) **(Fig 3.47)**

A tall (2-4 m), stiff, hard culmed marshy grass. *Leaves*, 1-5 cm wide, spreading pennant like; *Inflorescence*, a panicle, 25-50 cm long, feathery, spikelets with long silky hairs; *Ligule*, a ciliate line; *Rootstocks*, stout, often with long creeping stems; *Propagation*, by numerous seeds and rootstocks.

The plant is weedy in drainage ditches, shorelines, and waterlogged areas. In canals, rivers, and large lakes it starts on their shores but spreads fast with its creeping rootstocks, both inside the main waters and on the banks (even the lined ones). *P. australis* (Cav.) Trin. ex. Steud. (= *P. communis* Trin.) is another similar grassy weed.

41.2 *Saccharum bengalense* Retz. (*dharbi, kans*)

A robust, perennial, amphibious grass, about 1.5-2.5 m tall. *Leaves*, linear, usually over 1 m long and 1-2 cm wide, somewhat scabrous, sheath tight and hairy on margins above; *Inflorescence*, a panicle, 25-45 cm long, white, villus, spikelets lanceolate, acute, in pairs with one sessile and one pedicelled, the former with hermaphrodite florets and the latter with male or barren florets.

A troublesome grass of moist zone of banks of canals, rivers, and like situations.

41.3 *Echinochloa* spp. (barnyardgrass, jungle rice) **(Fig. 3.48)**

These are annual grasses, problematic in lowland rice fields throughout the country. In vegetative phase these resemble rice plants. *Stems*, up to 1 m tall, ascending from a creeping base; *Leaves*, coarse; *Inflorescence*, a panicle with bristled spikelets, often also awned. Two common species of this genus are *E. colonum* (Linn.) Link and *E. crus-galli* (L.) Beauv.

Paspalum spp. (knotgrass), *Brachiaria ramosa* (L). Stapf. (signal-grass), and *Phalaris minor* Retz. (reed-canarygrass) are semiaquatic grasses which often choke the drainage ditches, infest paddy fields, and abound the wet zone of the banks, depending upon the season.

42.0 Sedges

Family: Cyperaceae

Apparently grass-like herbs but characterised by their unbranched, triangular, solid stems, 3-rank, narrow leaves with closed sheaths, absence of ligule, flowers borne

Figure 3.47: *Phragmites karka* (Retz.) Trin. Ex. Steud. (common reed) (Biotype-4)

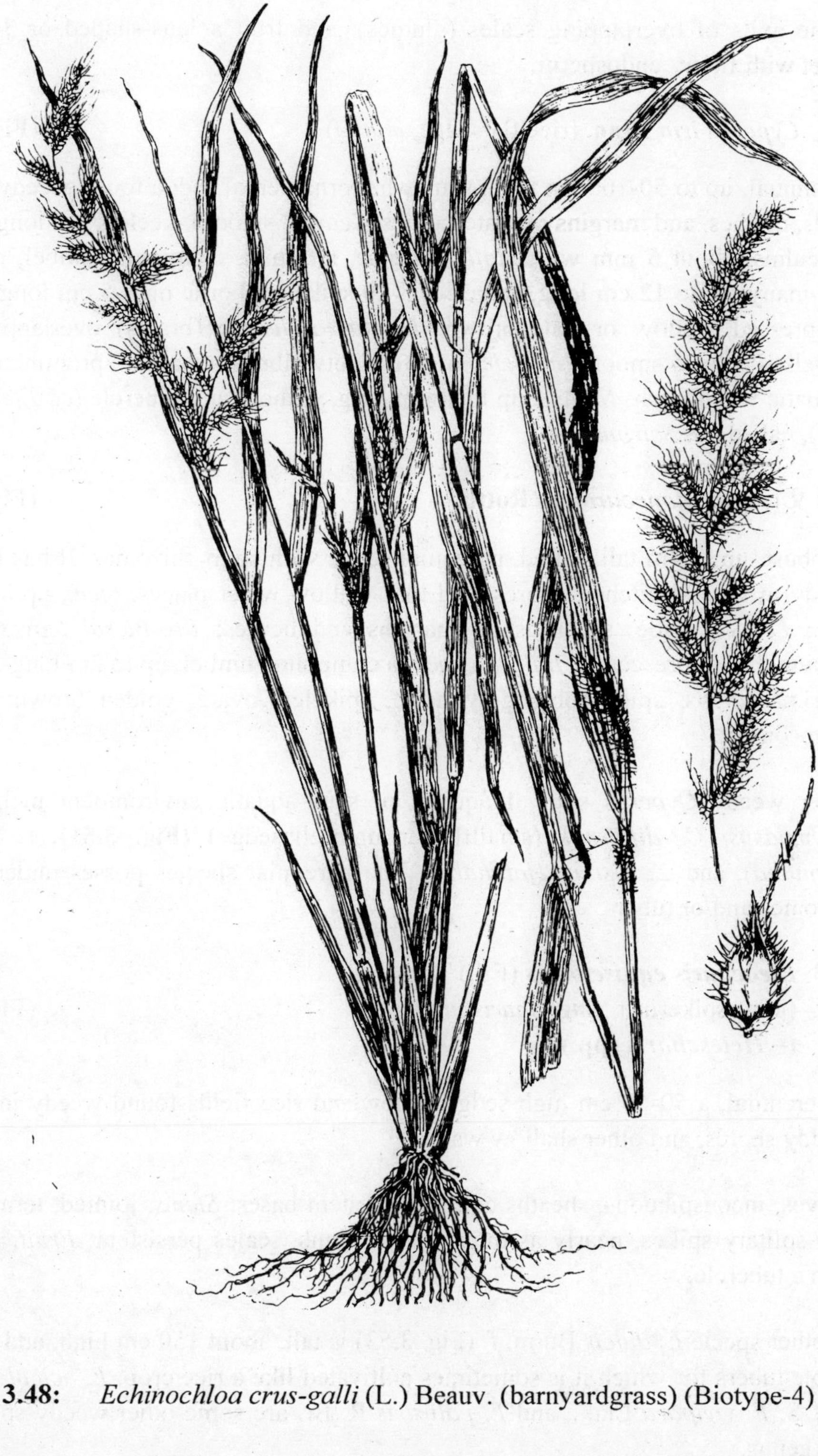

Figure 3.48: *Echinochloa crus-galli* (L.) Beauv. (barnyardgrass) (Biotype-4)

in the axils of overlapping scales (glumes), and fruit a lens-shaped or 3-angled nutlet with floury endosperm.

42.1 *Cyperus iria* Linn. (rice flatsedge, *mutha*) **(Fig. 3.49)**

An annual, up to 50-100 cm high, somewhat ornamental sedge found weedy in rice fields, ditches, and margins of waterbodies. *Leaves*, smooth, keeled, not longer than the culms, about 5 mm wide; *Inflorescence*, terminal, compound umbel, primary rays many, up to 12 cm long, spikes of 5-20 spikelets-borne on 1-5 cm long rachis, compressed, yellow or pale brown; *Involucral leaves* (bracts) overlapping the umbels; *Perianth* smooth (*cf. Scirpus* sp.); Roots, fibrous, shallow, prominently red, aromatic when fresh; *Nutlets*, up to 2 mm long, without any tubercle (*cf. Eleocharis* spp.), naked (*cf. Scirpus* spp.).

42.2 *Cyperus alopecuroides* Rottb. **(Fig. 3.50)**

A robust, up to 4m tall, tufted, perennial sedge with short rhizomes. It has become weedy in several ditches, shores, and like shallow water places. *Stem*, up to 1.5 m high; *Leaves,* wide, scabrous on margins and nerves; *Involucral leaves*, over-topping the inflorescence; *Inflorescence*, a compound umbel, up to 2m long with 5-7 primary rays, spikes oblong-cylindric, spikelets ovate, golden brown; *Nutlet*, planoconvex.

Other weedy *Cyperus* spp. of aquatic or semi-aquatic environment include, *C. corymbosus, C. difformis* (smallflower umbrellasedge) **(Fig. 3.51),** *C. dilutus* (*jalmutha*), and *C. distans* (*jalmutha*). The perennial species posses underground rhizomes and/or tubers.

42.3 *Eleocharis equisetoides* (Ell.) Torr.
(joint-spikerush, *singhapuri singhara*) **(Fig. 3.52)**
(= *Helescharis* spp.)

A perennial, a 50-60 cm high sedge of lowland rice fields found weedy in ponds, muddy shores, and other shallow waters.

Leaves, inconspicuous, sheaths confined to stem bases; *Stems*, jointed, terminating into solitary spikes, nearly as thick as the stems, scales persistent; *Fruit*, a nutlet with a tubercle.

Another specie *E. dulcis* Burm. f. (Fig. 3.53) is tall, about 150 cm high, and it bears edible tubers for which it is sometimes cultivated like a rice crop. *E. acicularis* (L.) R & S. *E. vivipara* Link., and *E. palustris* R. Br. are some other weedy species in this genus.

Figure 3.49: *Cyperus iria* Linn. (rice flat sedge) (Biotype-4)

Figure 3.50: *Cyperus alopecuroides* Rettb. (Biotype-4)

Figure 3.51: *Cyperus difformis* (smallflower umbrellasedge) (Biotype-4)

Figure 3.52: *Eleocharis equisetoides* (Ell.) Torr. (joint-spikerush) (Biotype-2)

Figure 3.53: *Eleocharis dulcis* Burm. f. (Biotype-2)

42.4 *Fimbristylis miliacea* (L.) Vahl (*bara jawane, joria*) **(Fig. 3.54)**

An annual or perennial, 40-60 cm high, tufted sedge, found weedy in largely lowland rice fields and shallow waters. *Involucral leaves*, shorter than the inflorescence; *Inflorescence*, an up to 10 cm long panicle, spikelets cylindrical with only lowest scale bearing flowers and fruits; *Fruit*, a nutlet (without tubercle).

Another weedy specie, *F. dichotoma* (L.) Vahl (*Jawane*) possesses involucral leaves which are longer than the inflorescence. It is better adapted to lowland rice fields than *F. miliacea.* *

42.5 *Scirpus* spp. (bulrushes) **(Fig. 3.55)**

Annual or perennial, rooted, emerged sedges of shallow ponds, drains, shores, and wet ditchbanks. Once established, these can withstand deep waters.

Stems, 1 m or more tall, *Leaves*, largely inconspicuous, sheaths present only at stem bases; *Inflorescence*, nearly sessile, cone-like, with silky bristles, terminal (although in species with single involucral leaf rising above the cone, it may look axillary from a distance), lowest scale of spikelets alone bearing flowers and fruits.

The common weedy species reported are *S. acutus* Muhl. (hardstem bulrush, big bulrush), *S. articulatus* (*chicchora*), *S. grossus* Linn. (*keseru*), *S. lacustris* (clubrush), *S. supinus, S. tuberosus* (*murek*, sea-clubrush), and *S. littorallis* Scharad.

42.6 *Cladium jamaicense* Crantz. (sawgrass, twig rush) **(Fig. 3.56)**

A perennial, 2-3 m tall, coarse sedge growing from stout rootstocks. A weed of shallow ponds, ditches, swamps, and wet ditchtanks.

Leaves, all basal, serrated saw-like; *Inflorescence*, a panicle, each branch terminating into a cluster of spikelets, turning brown towards maturity, spikelets cylindrical; *Fruit*, a nutlet (without tubercle) borne in the uppermost scale which alone bears flowers (*cf. Fimbristylis* and *Scirpus* spp.)

Note: *Involucral leaves: A cluster of leafy bracts in an inflorescence, usually at the base of a head of flowers.*
Tubercle: A swollen appendage present especially on the nutlets of some sedges.

Figure 3.54: *Fimbristylis miliacea* (L.) Vahl (Biotype-2)

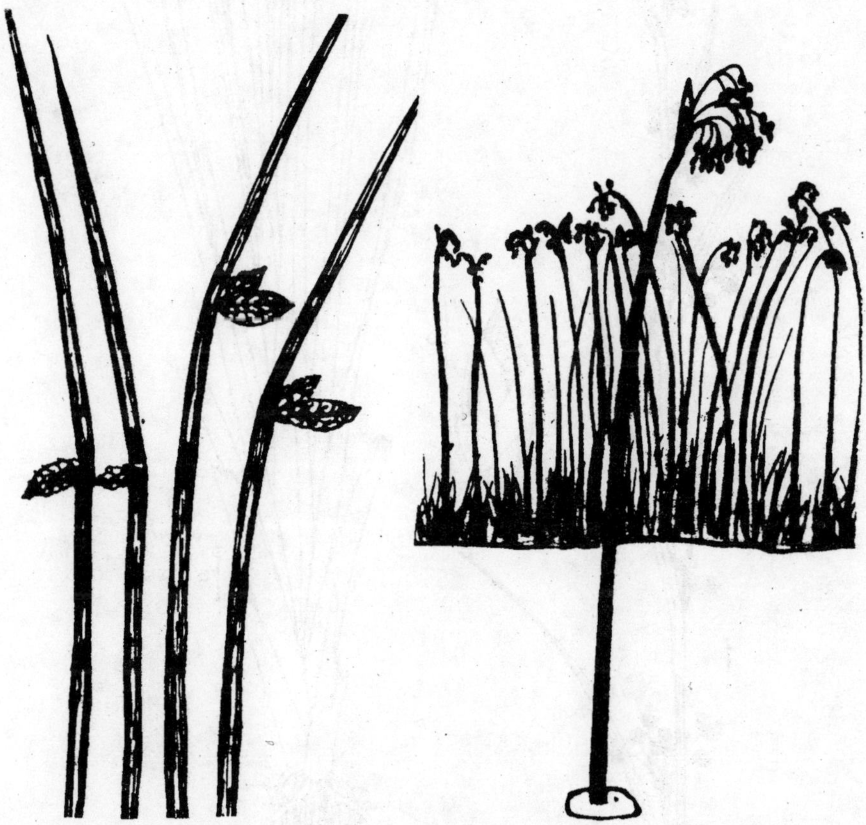

Figure 3.55: (Left), *Scirpus debilis* Pursh. and (Right), *S. validus* Vahl. (Biotype-4)

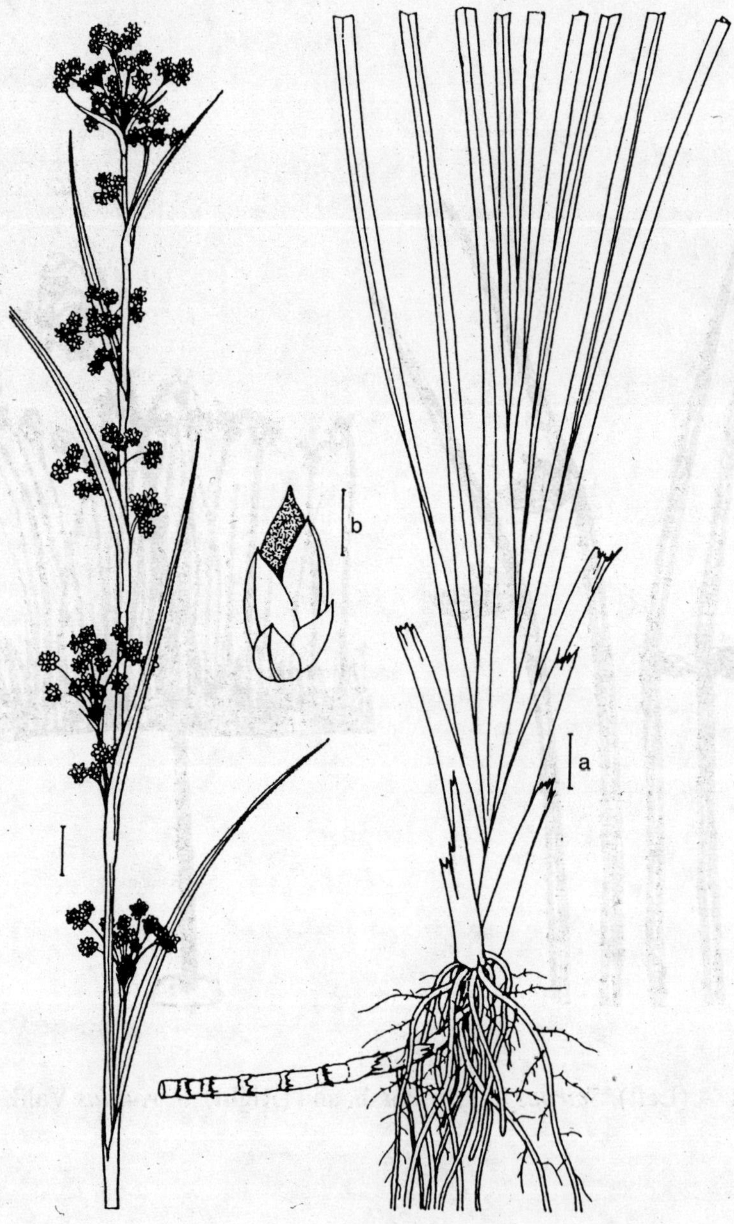

Figure 3.56: *Cladium mariscus* (L.) Pohl (Biotype-4)

42.7 *Carex* spp. (bullsedges) **(Fig. 3.57)**

These sedges are recorded weedy in shallow ponds, ditches, shores, and lowland paddy fields.

Leaves long, wavy; *Inflorescence*, of axillary unisexual spikes, usually the staminate spikes above and the pistillate ones below on the same shoot but, sometimes, each spike with mixed, unisexual flowers; *Fruit*, a nutlet enclosed in perigynium. *C. lasiocarpa* Ehrh. common weedy specie reported in this genus.

43.0 *Typha* spp. (cattails, *hogla, aiyra, patera*)
Family: Typhaceae

Robust, perennial, monocot, sedge-like, erect plants. *Stems*, stout or slender, nodeless, spongy; *Leaves*, long, linear, sheathing, without ligules, *Flowers*, dense, sessile, borne on cylindric, cushiony, terminal spikes in two portions, the staminate flowers in the above portion and the pistillate ones in the lower portion; *Propagation*, by profuse, hairy, wind-blown seeds (fruits) and extensive, underground, cylindric, yellow rhizomes. Rhizomes of a few cattail plants can occupy easily an acre of waterlogged land.

43.1 *T. latifolia* L. (common-cattail) **(Fig. 3.58)**

A 1-3 m tall, robust plant. *Stems*, stout; *Leaves*, linear, flat, 0.6-2.3 cm wide; *Spikes*, 15-30 cm long, the staminate and pistillate portions contiguous, the latter being dark-brown with black markings. It crosses freely in nature with other *Typha* spp. to give rise to hybrids.

43.2 *T. angustata* Bory & Chaub. (narrowleaf-cattail) **(Fig. 3.59)**
(= *T. angustifolia* L.)

A 1-1.5 m tall, cattail plant with creeping rhizomes. *Leaves*, narrow, up to 0.7 cm wide, somewhat rounded at the back; *Spikes*, with interval of atleast 0.5 cm between the staminate and pistillate portions, the latter being light or cinnamon brown, without black markings. *Propagation* and *Habitat*, similar to *T. latifolia*.

43.3 *T. elephantina* Roxb. (broadleaf-cattail, *mothitrina*, bora)

A gregarious cattail, up to 3.5 m high, with thick, stoloniferous rhizomes. *Leaves*, long, spongy, trigonous, deep green, very smooth, up to 3.8 cm wide; *Spikes*, with longer staminate portion than the lower pistillate portion; *Propagation* and *Habitat*, similar to *T. latifolia*.

Figure 3.57: *Carex lasiocarpus* Ehrh. (Biotype-4)

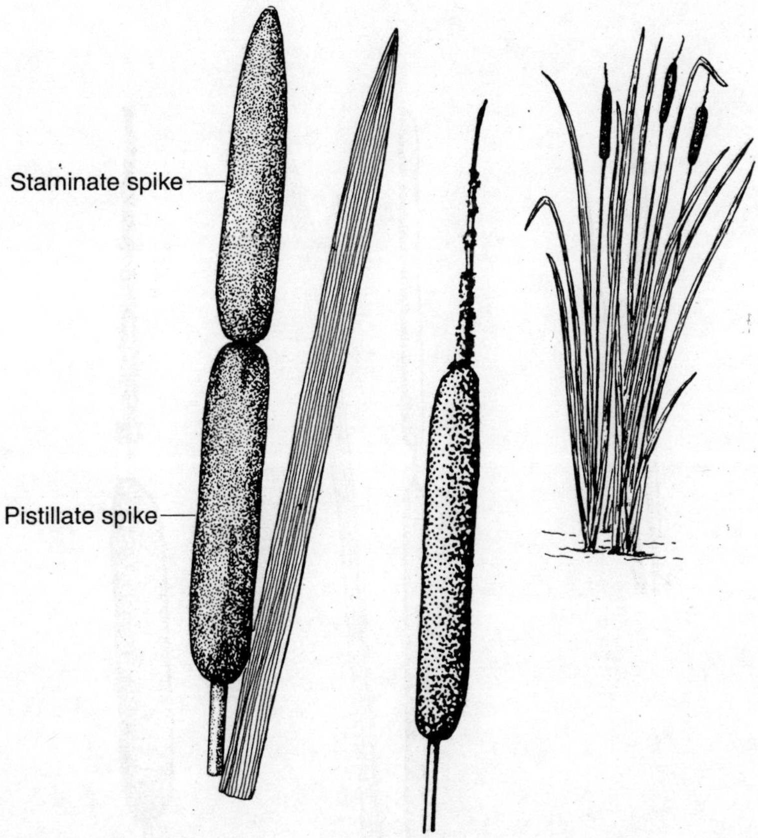

Figure 3.58: *Typha latifolia* L. (common cattail) (Biotype-4)

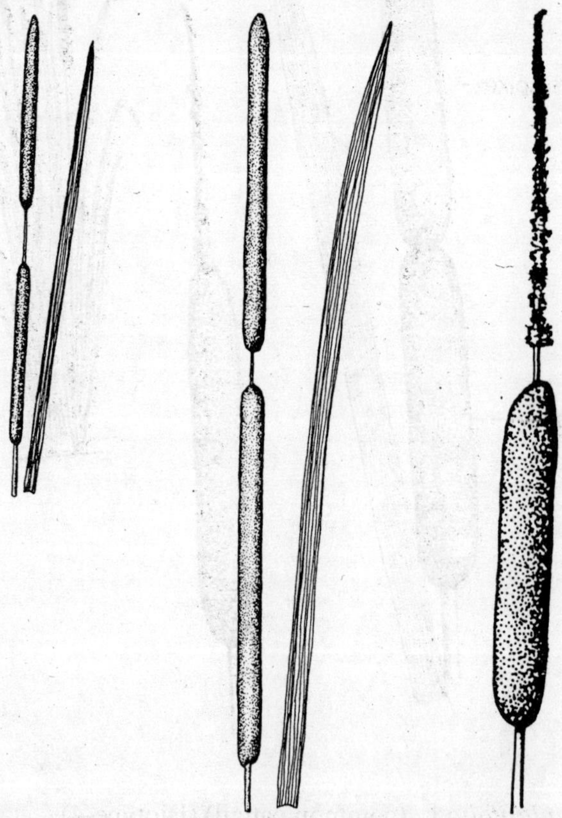

Figure 3.59: *Typha angustata* Bory & Chaub. (narrowleaf cattail) (Biotype-4)

SELECTED BIBLIOGRAPHY

19. Anonymous, 1970. Selected Weeds of United States. Agriculture Handbook No. 366, USDA, USA, 463 p.

20. Anonymous, 1992. Remedial Measures Project (Phase-1) for Indra Gandhi Main Canal, Rajasthan (India), GOR, Indra Gandhi Nahar Board, VII 239 p.

21. Cook, D.K. *et al.* 1974. Water Plants of The World. Dr. W. Junk b.v., Publishers, The Hangue, 561 p.

22. Crow, G.E. and C.B. Hellquist. 1981-83. Aquatic vascular plants of New England. Agr. Expt. Sta., Univ., Durham, New Hampshire Bulletin Nos. 517, 520 and 524.

23. Fassett, N.C. 1992. A manual of Aquatic Plants. Agro. Botanical Pub., Bikaner (India), 382 p.

24. Gupta, O.P. 1976. Aquatic weeds and their control. FAO Plant Prot. Bull. 24 (3): 76-82.

25. Haslam, M. 1978. River plants–the macrophytic vegetation of water courses. Cambridge University Press, Cambridge, 396 p.

26. Hellquist, C.B. and G.E. Crow. 1980-84. Aquatic vascular plants of New England. Agr. Expt. St., Univ. of New Hampshire, Durham, Bull Nos. 551, 518, 523, and 527.

27. Kaul, V. and D.P. Zutshi. 1967. A study of aquatic and marshland vegetation of Srinagar. Proc. Nat. Inst. Sci. India 33: 111-127.

28. Lange, L.D.E. *et al.* 1981. The occurrence of mixed populations of different genotypes of the *Lemna gibba* and *L. minor* complex. Acta. Bot. Neerl. 30 (3): 191-197.

29. Maheswari, J.K. 1965. Alligatorweed in Indian lakes. Nature (London), 206 (49-90): 1270.

30. Mirashi, M.V. 1954. Studies on the hydrophytes of Nagpur. J. Ind. Bot. Soc. 36: 306.

31. Mirashi, M.V. 1957. Studies on the hydrophytes of Umred. J. Ind. Bot. Soc. 39: 306-307.

32. Mitra, E. 1955. Contributions to our knowledge of Indian freshwater plants-2. *Limnathemum indicum* Thw. Proc. National Instt. of Sci. (India) 21 E (4): 170-178.

33. Mitra, E. 1964. Contributions to our knowledge of Indian freshwater plants-4. *Hydrilla verticillata* Royle. J. Asiat. Soc. 6 (12): 17-27.

34. Mitra, E. 1966. Contributions to our knowledge of Indian freshwater plants-5. *Pistia stratiotes*. J. Asiat. Soc. 8 (2): 115-133.

35. Nair, N.C. 1978. Flora of Punjab Plains. Bot. Surv. India, Indian Botanical Gardens, Howrah, 326 p.

36. Oommachan, M. and J.L. Shirivastava. 1997. Flora of Jabalpur. Scientific Pub., Jodhpur (India), 355 p.

37. Pattaniak, H. and C.M.K. Patnaik. 1956. The hydrophytes of Cuttack. J. Ind. Bot. Soc. 35 (2): 157-170.

38. Prasanna, V.R. and K.K. Rao. 1962. Algal resources of Pamban area. Ind. J. Fish 89 (1): 205-211.

39. Ramaprabhu, T. 1972. Observations on the autoecology of *Ceratophyllum demursum* with notes on its control. J. Asiat. Soc. 14 (2-4): 149-162.

40. Samant, S.S. and Y.P.S. Pangtey. 1994. Aquatic, marshy, and wetland flora of Pithorgarh (Kumaun, India). J. Eco. Taxonomy Bot. 18 (1): 47-54.

41. Sebastine, K.M. *et al.* 1960. Observations in the flora of Narasapur (Andhra Pradesh-India). Abstr. Proc. Ind. Sci. Congr. Pt. III: 406.

42. Smith, G.M. 1994. Crypotgamic Botany-Algae and Fungi. Tata McGraw-Hill Pub. Co. New Delhi, 546 p.

43. Stodola, J. 1967. Encyclopedia of Water Plants. T.F.H. Publications, Inc., N. Jersey, USA, 368 p.

44. Subramanyan, K. 1962. Aquatic angiosperms. Bot. Monograph No. 34, CSIR, N. Delhi, 190 p.

45. Yeo, R.R. *et al.* 1984. The morphology of hydrilla (*Hydrilla verticillata*) (L.F.) Royle). J. Aqua. Plant Manag. 22: 1-17.7

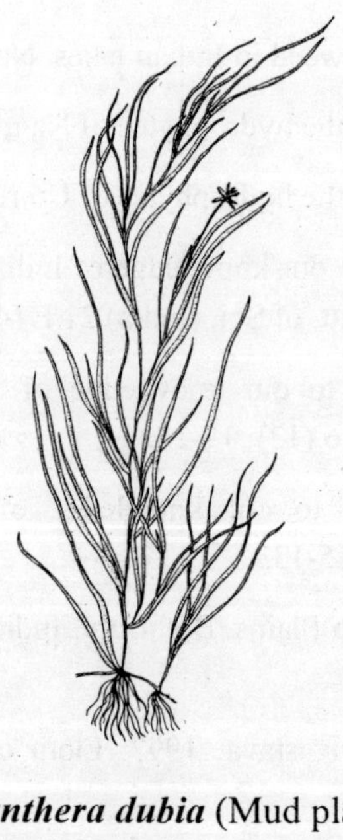

Heteranthera dubia (Mud plantain)

Chapter 4

SURVEY OF PROBLEM AQUATIC PLANTS IN INDIA

In India over 140 aquatic plant species are reported to have attained the status of aquatic weeds in different situations. Of these, the following ten species are of immediate concern to us. These are listed here *by alphabet* and *not* by their any kind of relative importance.

1. *Eichhornia crassipes* (waterhyacinth)
2. *Hydrilla verticillata* (hydrilla)
3. *Ipomoea carnea* (gramophone flower)
4. *Nelumbo nucifera* (lotus)
5. *Nymphaea stellata* (water-lily)
6. *Pistia* spp. (water lettuce)
7. *Potamogeton* spp. (pondweeds)
8. *Salvinia* spp. (waterfern)
9. *Typha* spp. (cattails)
10. *Vallisneria spiralis* (eelgrass).

Incidentally, the above list of primary aquatic weeds of India also includes three primary aquatic weeds of the world, *viz.*, *Eichhornia*, *Hydrilla*, and *Salvinia* spp.

Of the about 8 lakh ha of fresh waterbodies available in India for pisciculture, about 40% are rendered unproductive because of aquatic weeds. Some of the well known fishery lakes that have completely or largely yielded to aquatic weeds in the country are: Kitham lake, Barwar lake, Ramgarh lake, and Gujar lake in Uttar Pradesh, Kolleru lake in Andhra Pradesh, Ansupa lake in Orissa, Lotak lake in Manipur, Ooty lake in Tamil Nadu, and the world famous Dal, Nagin, and Wular lakes in Kashmir. In Karnataka, most of the fishery tanks in and around Bangalore and other cities have been badly invaded by waterhyacinth (*E. crassipes*).

Weeds have badly invaded our canal systems which are perhaps the largest in the world and made at enormous costs. The weeds have already reduced the designed flow of many of these canals by 40-50% (Gupta, 1973). In some cases, 80% reduction in water-flow due to aquatic weeds has been reported. The fortune costing Bhakra-Nangal canals in Haryana and Punjab, and the Chambal canals in Rajasthan and Madhya Pradesh are greatly slowed by the rampant growth of submersed weeds. Associated with the impeded flow of water in canals are miseries of forced seepage, waterlogging, and soil salinity in the command areas. The whole ecosystem around canals is thus rendered unfavourable for the natives because of the aquatic weed infestations.

Several irrigation and hydroelectric projects in the country are endangered by infestation of dams and reservoirs with massive growth of aquatic weeds. For instance, Tungabhadra project in Karnataka, Nagarjuna Sagar project in Andhra Pradesh, and Kakki and Idikki reservoirs in Kerala are filled with aquatic weeds to an alarming extent.

In the Brahmaputra river navigation is hindered by waterhyacinth while in the holy Mandakini river of Uttar Pradesh bathing is made unpleasant by several submersed weeds.

Vast areas of lowland paddy fields in the North-Eastern parts of India and Kerala state are badly infested with aquatic weeds. While in the North-East, *E. crassipes*, *Chara* and *Nittela* spp., and algal scums invade these fields, in the coastal Kerala *S. molesta* play havoc. Irrigation supply to paddy is hindered in about 1.6 lakh ha area in north-Eastern India alone. Add to this several hectares of cultivable flood–plains that are surrendered to noxious aquatic vegetation. Cultivation of waterchestnut (*Trapa* sp.) is also abandoned in East India because of heavy growth of waterhyacinth and other aquatic weeds in waterbodies.

In India certain fevers are transmitted by mosquito vectors that breed among specific submersed and floating weeds (Chapter-1). In potable water reservoirs, floating weeds like *S. auriculata* harbour snail hosts of urinary and intestinal disease organisms like *Schistosoma* spp.

The Bhilai plant in Madhya Pradesh and the Bokaro Steel Plant in Bihar are faced with acute problems of aquatic weeds in their cooling tanks where they prevent proper circulation of water in the pipes.

4.1: STATE-WISE AQUATIC WEED PROBLEMS

Now almost every state in India has aquatic weed problems but in certain states these are particularly serious. A brief survey of aquatic weed problems in different states of India is presented here as under. The states are arranged in alphabetical order.

1. Andhra Pradesh

Noxious aquatic vegetation is a serious concern to the freshwater fisheries and irrigation systems, besides the flooded rice fields, in Andhra Pradesh.

In the fishery lakes and tanks of the state, aquatic weeds are estimated to have already infested atleast 37% of these waterbodies. The major weeds there are: *Eichhornia, Azolla, Nymphaea, Nymphoides, Hydrilla, Vallisneria, Potamogeton, Najas, Euryale, Myriophyllum, Ceratophyllum, Utricularia, and Najas* spp., besides the waterbloom forming algae (*Microcystis* spp.) and the algal scums (*Cladophora, Ulothrix*, and *Spirogyra* spp.). *E. crassipes* has also increased incidence of filaria disease in the Godavari delta of the state. The famous Kolleru lake in West Godavari has yielded to *E. crassipes*, *Ipomoea aquatica*, and *Typha spp.*, besides some of the above cited weeds and algae.

In irrigation systems of the state, *I. aquatica* has blocked the drainage channels, causing silting and floods. *I. carnea* has gone wild throughout the irrigation system in Andhra Pradesh, planted initially on banks as a soil binder and in fisheries as a shelter plant. In the Nagarjuna Sagar Project on the other hand, *T. angustata* is the main bank weed. The main drains of Krishna and Godavari are choked with *Ipomoea* spp. and *E. crassipes*, resulting in submergence of the adjoining rice fields to great sorrows of the farmers there.

In rice fields, *Ottelia alismoides*, *Monocharia vaginalis*, *Marsilea quadrifoliata*, and species of *Cyperus, Eleocharis, Scirpus, Spirogyra, Chara*, and *Nitella* are additionally problematic. In recent years, *Pistia stratiotes* has also spread in the waterways.

2. Assam

Aquatic weeds are relatively common in the plains of Assam in her Borak and Brahmaputra valleys where both natural and man-made waterbodies are badly infested. Aquatic weeds make them less fit or even unfit for fish culture and irrigation. Most irrigation water tanks show 86-92% weed infestations. Major aquatic weeds present in varied waterbodies of Assam are: *E. crassipes, Pistia stratiotes, Salvinia* sp., and *Lemna minor* amongst the free-floating biological types; *Sagittaria guayanensis* sub-sp. *lappula, Nymphoides indica*, and *Nymphaea* sp. amongst the anchored floating types; *S. sagittifolia, Monochoria vaginalis, M. hastata, Scirpus juncoides, Eleocharis dulcis*, and *E. acutangulus* amongst the anchored emersed types; *Hydrilla verticillata* and *Vallisneria* sp. amongst the submersed types; and *Utricularia bifida* and *Chara* sp. amongst the suspended biological types. Besides, in the flooded rice fields, *Alternanthera philoxeriodes, Leersia hexandra, Marsilea qudrifolia, Oryza rufipogon, Hymanachne acutigluma* (Steud) Gil,* and *Spacciolepsis interrupta* (Willd). Stapf*, are reported weedy. Also, many drainage ditches are choked badly with *Colocasia antiquorum*.

3. Haryana

Typha angustata and *Eichhornia crassipes* are serious aquatic weeds in Haryana. In the Canal Command Area the drainage system is badly choked with these two weeds, besides with *Paspalum distichum* (knotgrass) in different locations. In the flood areas of western part of Haryana, *Typha* has covered thousands of hectares of culturable land. This robust weed has also become a serious menace in roadside ditches where the rain water stagnates every year. Similar miserable situation exists all along the Jamuna and Bhakra canals.*

In Western Haryana, the village ponds are frequently invaded by waterhyacinth brought with floods and it disseminates from one pond to another, pretty fast.

In the holy Brahm Sarovar lake in Kurukshestra, aquatic weeds, mainly, *Hydrilla, Vallisneria,* and *Ceratophyllum* spp., have spoiled its water quality for bathing, recreation, and potable uses. Also, the shore waters support heavy growth of scum algae which makes the stone staircases slippery, besides emitting foul smell in the environment.

4. Himachal Pradesh

In Himachal Pradesh aquatic weeds are problematic in her irrigation water channels fed from wells and storage tanks in the lower hills and in irrigation streams in the mid-hills receiving water from the upper hills. Here a variety of submersed and emersed aquatic weeds retard the flow of water and many a time choke these completely. In fish ponds located in the low and mid-hills, fish yields are greatly reduced at the hands of variety of aquatic weeds. In the famous, natural lakes Prashar, Rewalsar, and Renuka, aquatic weeds mar their aesthetic values.

---

* * *Family = Poaceae*

i. **Submersed species:** *Potamogeton nodosus*, *P. crispus*, *Najas marina*, *Hydrilla verticallata*, *Ranunculus* sp., *Utricularia vulgaris*, *Eleocharis acicularis*, *E. elegans* and *Marsilea minuta* (a fern). *Chara* and *Spirogyra* spp. are the main algae present.

ii. **Surface species:** *Nastrutium microphyllum* and *Salvinia auriculata*.

iii. **Erect species:** *Echinodorus trialatus*, *Scirpus* sp., *Juncus* sp., *Sagittaria* sp., *Cyperus* spp., *Plantago major*, and *Ponteaderia cordata*.

Besides the above, in the palins of Himachal Pradesh, bordering Haryana and Punjab, *Eichhornia crassipes*, *Typha* sp. and *Phyragmites communis* invade various waterbodies and wetlands.

5. Jammu and Kashmir

The world famous beauty lakes of Kashmir, namely the Dal, Nagin, and Wular are sick with aquatic weeds. These lakes attract thousands of tourists each year as grand venues of aquatic sports, including skiing. But now these sports are at stake because of invasion by aquatic weeds. Besides, the efficiency of irrigation and pisciculture is greatly impaired by aquatic weeds. The aquatic weed flora in these lakes varies with altitude. Among the floating weeds, the common species are *Eichhornia crassipes*, *Nymphoides peltata, Polygonum amphibium, Nelumbo nucifera, Nymphaea* spp., *Trapa natans, Lemna gibba, L. minor, L. trisulca, Spirodela polyrhiza,* and *Salvinia natans*. The most dominant submersed weeds in the lakes are: *M. spicatum, Ceratophyllum demersum*, and *Potamogeton* spp. *C. demersum* infests largely the deeper portions of the lakes.

The side channels of lakes used for transporting irrigation water are choked with *Hydrilla* and *Potamogeton* spp. In the lowland rice fields *Potamogeton* and *Chara* spp. pose problems in Kashmir.

Of the several aquatic weeds, *S. natans* has been so far reported weedy in India only from Kashmir. It covers 2.5-6% water surface of most lakes there, though in Gilsar lake the fern coverage exceeds 30%. The fern is now heading fast into the rice fields.

On banks of the lakes *Typha angustata* and *Phragmites communis* are very common. A number of other emergent aquatic weeds like *Butomus, Juncus, Carex, Scirpus, Sagittaria*, and *Alisma* spp. form dense stands in the slow streams and irrigation ditches where they obstruct water-flow and cause silting and floods. Many of such weeds escape into lowland paddy fields and reduce crop production.

6. Kerala

Salvinia molesta is the most notorious aquatic weed in Kerala. It is a serious threat to her hydroelectric projects, pisciculture, navigation, and lowland paddy fields. The fern was introduced into Kakki reservoir of Sabarigiri hydroelectric scheme in 1967 as packing material for a boat. By 1973 it had spread to 200 ha of water. Now it is also a problem in Idikki reservoir and in the coastal paddy fields. In 1977, *S. molesta* covered as much as 1900 ha of paddy land. Its young sporophytes adhere to the crop seedlings and grow rapidly into thick mats.

Salvinia is such a competitive plant that it has out-witted *E. crassipes* and *Pistia stratiotes*, the two historic aquatic weeds of Kerala. During tidal movements, especially in the back water areas and estuaries, large quantities of salvinia are transported to the inshore regions of the sea, and they pollute the beaches and intertidal zones, thus affecting the littoral fauna. The inshore benthic community is also affected to a great extent by the settlement of dead, decaying salvinia plants.

T. angustata is an important shoreline weed in Kerala. *Cyperus deformis, C. haspan, Fimbristylis dichotoma, Eleocharis dulcis*, and *Ludwigia lustris* are weedy in paddy fields, irrigation channels, and shorelines.

7. Maharashtra

In Maharashtra, aquatic weeds pose serious problems in her canal irrigation systems, including in the drainage channels and waterways. Potable water supply systems, pisciculture, and lowland rice fields are other aquatic environments seriously involved by the weeds.

In canals, *Potamogeton* and *Vallisneria* spp. dominate. They hinder the water flow and also force accelerated seepage to cause waterlogging all around in the deep black cotton soils of the command area. The drainage ditches and waterways are badly clogged with *Typha, Cyperus*, and *Ipomoea* spp. On ditchbanks, and in many places even within the ditches, the notorious *Parthenium hysterophorus* and *Prosopis* spp. dominate the scene and hinder free approaches to them, besides causing bank erosions.

In the potable water systems and pisciculture of the state, *V. spiralis*, *Hydrilla verticillata, Chara* spp. and various scum algae grow abundantly. The major potable water systems affected adversely by these plants are Kanhan river and several desilting tanks and storage reservoirs. *H. verticillata* harbours epiphytic algae too. Excessive chlorination of potable water in the reservoirs often adopted to counter the disease causing pollutants and organisms occurring there because of the presence of aquatic weeds and algae, imparts disagreeable odour to it. *V. spiralis,* in particular, offers breeding ground to several mosquito and snail vectors of specific human diseases.

In fishery water tanks like the Gandhinagar tank and Phutala tank, waterblooms, caused by chiefly *Microcystis* and *Spirulina* spp., cause heavy fish mortality.

In lowland and deep water paddies some other aquatic plants found weedy include: *Aeschnomena indica, Sphenochloa zeylanica, Jussiaea repens, Pistia stratiotes, Rungia repens* Nees., *R. parviflora* Nees., *Alisma plantago-aquatica, Scirpus acutus, S. grossus, S. articulatus, Cyperus iria, Lemna minor, L. trisulca, Limnophila heterophylla, Aponogeton natans*, and *Utricularia* sp.

8. Orissa

In Orissa the coastal districts of Puri, Cuttack, Balasore, and some parts of Ganjam are worst hit by aquatic weeds. These areas have many branches of rivers and several canals, besides being famous for their round the year cultivation of flooded rice. All these conditions force large scale seepage and waterlogging, favouring heavy weed growths. The aquatic weeds and their allies have turned large cultivable lands into wasted lands. While *E. crassipes* occupies 75% area of the coastal district waterbodies, a number of other aquatic weeds like, *Salvinia cuculata, S. natans, O. alismoides, N. stellata, Nymphoides peltatum, N. indica, I. carnea, I. aquatica, V. spiralis, Trapa natans, Najas indica, N. graminea, N. verticillata, Polygonum tomentosum, Nechamandra alternifolia, P. nodosus, C. demursum, Wolffia arrhiza, T. angustata and Centella, Ludwigia, Marsilea, Chara, Spirodela* and *Microcystis* spp. also invade these waterbodies.

In the mainland of Orissa, aquatic weeds badly infest lakes, farm ponds, and public tanks which are very important sources of fish to the natives. In Ansupa lake, the largest freshwater body in the state, a variety of aquatic weeds cover about 85% the waterbody. *Hydrilla, Najas, Ceratophyllum, Ottelia, Vallisneria*, and *Chara* spp. are the more problematic weeds there. In the public tanks, weeds like *Nymphaea, Nelumbo*, and *Nymphoides* spp. are troublesome. In the derelict tanks, marginal weeds like *Jussiaea, Ipomoea, Eleocharis*, and *Marsilea* spp. are disturbing.

9. Rajasthan

As is true of many other states of India, water is scarce in Rajasthan too. In fact, in her North-Western region it is so scarce that women-folk have to walk a couple of km to fetch pitchers of water from the surface ponds. Before 1947, the state of Rajasthan had only one canal located in the erstwhile state of Bikaner. But subsequently the state has added several major and minor water projects. The major amongst these are: Indra Gandhi Canal, Beas Canal, Bhakra Canal, Chambal Canals, and the Mahi Canal systems. Besides, Rajasthan has over three thousand water storage tanks and several big artificial lakes. These lakes form important sources of water for fishing, irrigation, recreation, and for drinking water supplies. Some of these lakes offer great tourist attractions too.

Aquatic weeds pose prime problem in the maintenance of waterbodies in the state. The Chambal irrigation system, comprising two canals and their several distributaries (shared by Rajasthan and Madhya Pradesh), was commissioned in 1960. Survey conducted in 1975 revealed that aquatic weeds cut the designed discharge of canals by 40-70%, and that of their distributaries by over 80%. The major weeds there were: *Potamogeton, Hydrilla, Vallisneria*, and *Najas* spp. These weeds now occupy over 1500 ha of the canal area in the Chambal project. Besides reducing the irrigation water reaching the distal ends, the weeds have aggravated seepage losses of water, which in turn has created large waterlogged areas around. *Typha* spp. have invaded these wagerlogged areas in about 10-15 thousand ha land. Also, the drainage ditches and distributaries of chambal canals are filled with *Typha* spp. A new drain laid in the area is filled with *Typha* within about six months.

In the Indra Gandhi Canal Project (IGCP) located in the North-Western region of Rajasthan, *Typha* spp., *Phragmites karka,* and *Parthenium hysterophorus* have covered the entire canal shores and banks. *Typha* and *Phragmites* are now slowly heading from the shores into the main stream where *P. perfoliatus, P. nodosus, Vallisneria* spp., and *H. verticillata* are already present. The unlined branch canals and distributaries of the Indra Gandhi canal are invariably infested with weeds like *Typha, Phragmites, Saccharum, Polygonum, Ipomoea*, and *Nympoides* spp. In many parts of the canal now *E. crassipes* has also spread.

P. karka has invaded many portions of even the lined banks of the canal. Aquatic weeds have also invaded and slowed down the water flow in the canal distributaries and waterways by over 50%. Besides checking the waterflow, the aquatic weeds have impaired the quality of its water for drinking. The hyacinth mats also impede or prevent in many places the boat traffic in IGMC. Further, during the floods waterhyacinth mats block bridges, culverts, and fences, thereby raising the flood levels.

In several drinking water lakes and fish ponds in Rajasthan, *E. crassipes* is the major weed. Besides, *Trapa, Pistia, Nymphaea, Nymphoides*, and *Nelumbo* spp. cover the impounded waters. Submersed weeds like *Hydrilla, Vallisneria, Chara, Potamogeton*, and *Najas* spp. are very common in Rajasthan. In the lakes at Udaipur and Bharatpur aquatic weeds, particularly *E. crassipes* and *H. verticillata*, are potential dangers. In shallow recreational waterbodies, such as around the decorative fountains, scum algae (chiefly *Cladophora* and *Pithophora* spp.) offer repulsive sight and odour. In fish nursery tanks, blue-green algae form the deadly waterbloom.

10.Tamil Nandu

In Tamil Nadu aquatic weeds pose serious problems in and around almost all kinds of waterbodies, including rivers, canals, system tanks, reservoirs, lakes, and hydel power generation systems. In canals and their distributaries, aquatic weeds have already reduced in many places their carrying capacities by 70-80%. The two common notorious weeds responsible for this situation are *Eichhornia crassipes* and *Impomoea carnea.* Besides, the canals are infested with other aquatic weeds like *Hydrilla, Vasllisneria, Nelumbo,* and *Nymphaea* spp. In vast waterlogged areas created around canals, as also in their shorelines; *Typha spp.* are the main weeds.

In the famous rivers cauvery, Periyar-Vaigai, and Thamirabarani, *I. carnea* is the major bank and shore weed from where it has gone wild into the respective canals. Severe blockage of water by *I.carnea* has led to the formation of mini silt islands (isles) in the rivers, which now grow several weedy bushes and cause floods. These floods disseminate masses of waterhyacinth to nearby paddy fields, causing enormous losses in paddy production.

In several reservoirs, lakes, and system tanks in the state used for fishing, recreation, and supplying drinking water, a variety of aquatic weeds hinder or completely prevent their assigned uses. The prominent among these weeds are *Hydrilla, Vallisneria, Myriophyllum, Najas, Ceratophyllum, Potamogeton, Lemna, Eichornia, Salvinia, Pistia, Nelumbo, Nymphaea, Alternanthera, Typha, Monochoria, Ipomoea, Sagittaria, Fimbristylis,* and *Phragmites* spp. On banks of these waterbodies several terrestrial weeds hinder their free approaches.

In urban areas several water ponds used in the past for vegetable cultivation during the summer months are now abandoned due to invasion by waterhyacinth. Frequent disposal of city sewage water in them has led to the present rampant growth of *E. crassipes* there.

11. Uttar Pradesh

Some of the largest fishery lakes of Uttar Pradesh have come under the grip of noxious aquatic vegetation for varied reasons. In these lakes the fish production is now either less economical or uneconomical. In the 300 ha Kitham lake near Agra, for example, the annual fish production potential worth Rs. 2 lakhs was reduced to almost nil because of large coverage of its water by *E. crassipes*. The weed entered this lake through Okhla canal.

The 1000 ha Barwar lake in Jhansi is so heavily infested with submersed weeds that its potential annual fish revenue has been reduced to one fourth. *P. perfoliatus, P. crispus,* and *P. nodosus,* in association with *Hydrilla, Vallisneria, Najas*, and *Chara* spp., are the main culprits for causing this great economic loss to the state. In Gorakhpur, the 22 sq m Ramgarh lake has got filled with dense growth of *Hydrilla, Najas, Potamogeton, Ceratophyllum*, and *Chara* spp. Of these, *Hydrilla* and *Najas* spp. infest the lake round the year while others invade it seasonally. The overall weed biomass in the lake is at peak during December to August.

The 110 ha Gujar lake in Varanasi is another sad story of invasion by aquatic weeds. It is a shallow lake with only 1-3 m water depth in different parts of the year. In the middle of the lake, submersed weeds like *Hydrilla, Najas*, and *Potamogeton* spp. are dominant while in the shorelines, floating weeds like *Nelumbo, Nymphaea, Nymphoides*, and *Eichhornia* spp., are common. In its marsh zone, emergent weeds, chiefly *Oryza, Eleocharis, Scirpus*, and *Paspalum* spp. engage the ground.

In *Tarai* area of Uttar Pradesh the fishery ponds are also losing ground to aquatic weeds like *Hydrilla, Potamogeton, Vallisneria, Nelumbo*, and *Nymphaea* spp. Marginal weeds, chiefly *Typha, Saccharum*, and *Brachiaria* spp. hinder easy approaches to the waterbodies.

The aquatic weed problems in the state extend into rivers too. In Mandakini river at Chitrakoot the aquatic weeds have made holy bath for the piligrims cumbersome. The main weeds there are *Hydrilla, Vallisneria, Scirpus, Nymphoides, Potamogeton, Phragmites*, and *Cyperus* spp., besides the ugly and offensive algal scums. *T. angustata* abounds widely on river banks and in marshy areas in Uttar Pradesh.

12. West Bengal

E. crassipes is the foremost aquatic weed of West Bengal ever since early part of the 20th century. Vast mats of this weed are a common sight in the state in all kinds of waterbodies. Other common floating weeds in the state are: *Ipomoea, Pisita, Nymphaea*, and *Nelumbo* spp.. *Salvinia* and *Lemna* spp. are weedy in scattered areas. Of the submersed weeds, *Hydrilla* is the biggest nuisance. Among the emersed weeds, *Monochoria, Sagittaria, Polygonum, Cyperus*, and *Scirpus* spp. are common throughout the state in the shallow waterbodies. In the Southern part of West Bengal, *Typha* is a noxious weed. Aquatic weeds have played havoc in West Bengal in her fishery waters, potable waters, and lowland paddy fields. In Palta and Baranagar water works, *Eichhornia* and *Lemna* spp., alongwith some molluscs, constrict the water pipes. In Southern and Coastal West Bengal, *Chara* and *Nitella* spp. invade heavily the paddy fields where they impoverish crop growth and emit bad odour. In deep water paddy fields, *E. crassipes and Ipomoea* and *Scirpus* spp. cause extraordinary crop losses.

SELECTED BIBLIOGRAPHY

1. Anon. 1992, Remedial Measures Project (Phase-I) for Indra Gandhi Canal, Rajasthan, GOR, India.

2. Brezny, O. and I. Mehta. 1970. Aquatic weeds and their distribution in Chambal command area in Kota (Rajasthan, India). Ind. J. Weed Sci. 2 (1): 70-80.

3. Govindu, H.C. and Y.C. Panchal. 1970. Some common weeds of Mysore State II. Weeds of aquatic areas, marshylands, and drying ponds. Myforest 6 (4): 53-60.

4. Gupta, O.P. 1973. Aquatic weed control. World Crops 25 (4): 132-190.

5. Jana, S. and M.A. Choudhuri. 1983. Aquatic weed problems in West Bengal (India). Weeds Today 14 (1): 9.

6. Joshi, N.C. 1973. Some problems and progress of weed control in India, 1948-1972. Pesticides 7 (3). 9-12.

7. Kaul, V. and U. Bakaya. 1976. The noxious floating *Lemnid-Salvinia* aquatic weed complex in Kashmir. IN Aquatic weeds in S.E. Asia, Dr. W. Junk b.v. Publ., The Hague, 183-192.

8. Kaul, V. *et al.* 1976. Aquatic weeds in Kashmir. *IN* Aquatic Weeds in South-East Asia, Dr. W. Junk b.v. Publ. The Hague, 70-83 pp.

9. Majumdar, R.B. 1962. Weed flora of the district of 24 Parganas (West Bengal, India). Ind. Agric. 6 (1-2): 208-213.

10. Malhotra, S.P. 1976. Remedy for aquatic weeds in Bhakra canals. *IN* Aquatic Weeds in S.E. Asia, Dr. W. Junk b.v. Publishers, The Hague, 253-254 pp.

11. Mehta, I. 1979. Problem and control of *Typha* in the Chambal Command area. Ind. J. Weed Sci. 11 (1-2): 36-46.

12. Misra, G. and N. Das. 1969. Studies on the control of aquatic weeds in Orissa. Hyacinth Contr. J. 8: 22-23 and 40-41.

13. Mukhopadhyay, S.K. and S.K. Taraphdar. 1976. Aquatic weed problems in West Bengal and control through herbicides. *IN* Aquatic weeds in S.E. Asia, Dr. W. Junk b.v. Pub. The Hague, 269-276 pp.

14. Nair, A.K. and T.R. Rai. 1973. Kakki Reservoir--Environmental Effects on Water Weeds and Their Control. Hydro Power (published by Kerala State Electricity Board, India): 13-16 pp.

15. Ozimek, T. 1984. Long term changes of the submersed macrophytes in Eutrophic lake in North Poland. Aquat. Bot. 19: 1-11.

16. Patnaik, S. 1971. Seasonal abundance and distribution on bottom fauna of the Chilka lake. J. Marine Biol. (India) 13 (1): 106-125.

17. Raju, R.A. 1999. *Ipomoea* menace in Andhra Pradesh. Weed Sci. News Letter, 3 (1 & 12): 4-5.

18. Ramachandran, V. *et al.* 1976. A survey of aquatic weed infestation in Andhra Pradesh. *IN* Aquatic Weeds in S.E. Asia Dr. W. Junk b.v. Pub., The Hague, 91-98 pp.

19. Sankaran, T. and E. Naraynan. 1971. Occurrence of the alligatorweed in South India. Curr. Sci. 40 (23): 641.

20. Soerjani, M. *et al.* 1975 Aquatic weed problems and control in South East Asia. Hyacinth Control J. 13 (6): 2.

21. Tosh, G.C. and G.K. Patro. 1974. Survey of major aquatic weed flora in the coastal tracts of Orissa. JNKVV Research J. (India) 9 (1 & 2): 10-12.

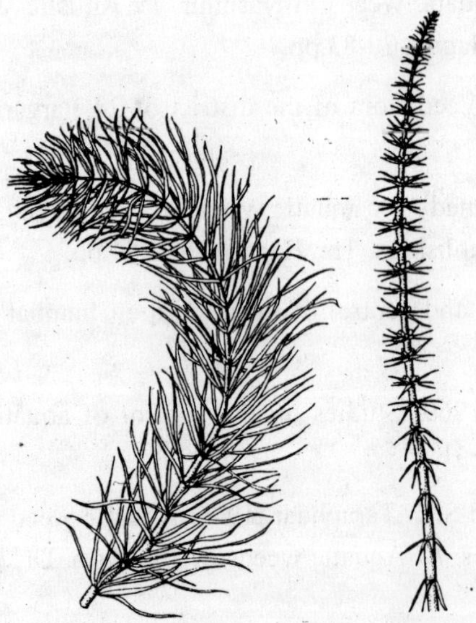

Hippuris vulgaris (Mares tail)

Chapter 5

AQUATIC PLANT MANAGEMENT STRATEGIES

Aquatic plant management is a tremendously costly operation, both initially and in terms of continued weed suppression. Countries conscious of this menace are spending millions of dollars each year on this programme to harvest benefits of weedfree waterbodies. As such, it is important to formulate appropriate strategies for achieving economic aquatic weed management, before finalising the action programme.

Aquatic plant management strategies typically have five stages:

(i) Identification and Assessment of the Problem.

(ii) Control Selection.

(iii) Control Implementation.

(iv) Long Term Management.

(v) Ecological Shifts.

5.1: IDENTIFICATION AND ASSESSMENT OF THE PROBLEM

Mere existence of some plants in or around a waterbody is not enough to justify their control. Before doing so, one must be able to enumerate the kind and extent of damage being inflicted by individual plant species to the utility of the waterbody. Besides, we should foresee the immediate and future gains and losses that might result if it was decided to destroy the present vegetation. One can then weigh them properly to come to conclusion about the degree to which the aquatic vegetation should really be controlled. In no case total vegetation kill must be aimed at. A certain amount of vegetation is almost always needed in water to replenish the dissolved oxygen, remove turbidity, and provide aesthetics. Likewise, on the ditchbanks some suitable vegetation to check soil erosion is necessary. Even if an aquatic weed control programme is found warranted in or around a waterbody, one must give further consideration to the resulting ecological changes, including shifts in plant species. However, a few strange looking plants with known potential of quick growth in or around a waterbody must be checked immediately, without long period considerations. *Parthenium hysterophorus*, for instance, is one such ditchbank weed that is capable of spreading like wild fire and it must be eliminated at the first appearance. So is the case with *Eichhornia crassipes* and *Salvinia* sp. inside new waterbodies.

Further, while we need to know how to control our weedy aquatic vegetation, we must simultaneously think of encouraging the growth of desirable plants, especially on the banks. A balanced growth of the right species of aquatic plants should be our goal in the management of any productive waterbody.

Many a time the size of a weed population is used as the main criterion of estimating the extent of the problem. It is so because usually the aquatic weed problem is recognised at a late stage when their damage is apparent. Also, once the present weeds are controlled, often further analysis of the cause of the problem looks irrelevant and control procedures tend to be repeated whenever necessary. But for a long term management of a waterbody, a full evaluation of the causes and nature of the weed problem need be studied, taking into account many variables. Stability of the hydrological regime, nutrient status, climatic conditions, growth and reproductive potential of the weedy plants, their effects on physicochemical and biotic factors in the environment and whether they are alien species, all need be considered. At the present there are no generally established standard procedures for assessing aquatic weed problems and predicting their seriousness, although these would be very desirable for monitoring the potential problems. Mitchell (1979) proposed a proforma to record assessment of aquatic weeds (Table 5.1). It can be modified to suit specific situations.

Table 5.1: Field Assessment Form for Recording Aquatic Weed Problems (Mitchell, 1979)

Assessment number: - Date:

Completed by: -

1. Site Identification And Description

   (Longitude, latitude, locality, visit number, etc.)

2. *Weed Identification And Occurrence*

   (Major weeds; alien or native)

3. *Nuisance Status*

   (Uses of the waterbody in order of importance; rated annual value; estimated annual loss due to weeds)

4. *Environmental Status*

   (Age and type of waterbody; stability of hydrological regime; light penetration into water; monthly mean temperatures; average days of frost per year, and nutrient status)

5. *Plant Status*

   (Period of infestation, dispersal, and introduction agency; area of infestation; growth rate; vigour; growth form; reproduction; risk of reinfestation).

6. *Place In Ecosystem*

   (Primary production, direct food source, invertebrate habitat, substrate for epiphytes, shelter for vertebrates, and competition with other non-weedy plants)

7. *Other Notes*

Experienced weed scientists will probably find little advantage in the use of the field assessment form. However, assessors are not always experienced, and management decisions may require that situations be monitored by means of respective, objective assessment procedures.

5.2: CONTROL SELECTION

Of the various techniques available to date for the management of aquatic weeds, both preventive and the direct control ones, local conditions must dictate the choice amongst these. Haste in adopting any new weed control method must be avoided. One must properly evaluate all alternative control strategies and then formulate an appropriate management programme. One has to consider in this exercise several aspects as under: -

1. Present and future possible uses of the waterbody including the fate of the outgoing water.

2. Objectives of controlling the aquatic weeds and criteria of evaluating the effectiveness of the technique in meeting these objectives, *e.g.* gains in quantity and quality of

aquaculture, flow rates of streams, protection of hydroelectric plants, aesthetic values of the waterbody, aquatic sport possibilities, ease of propelling, etc.

3. Cost-benefit and cost-effectiveness analyses of alternative systems. This could include both fixed and recurring costs and effectiveness.
4. Impact of the alternative methods on the surrounding ecosystem.
5. Ease of adoption of the practice in relation to local availability of inputs and technical manpower.
6. The kind, age, and extent of weed infestation.
7. Accessibility to the waterbody by machines of different sizes and operating nature.
8. The depth of water, possible seasonal fluctuations, and its human regulatability.
9. Rapidity and kinds of possible ecological shifts.

Mitchell (1979 a) has proposed several evaluation components of treatment methods of aquatic weeds to weigh their benefits and costs. These are summarised in table-5.2.

Table 5.2: Evaluation Components of Aquatic Weed Infestations and Their Management (Mitchell, 1979 a).

1. *Annual value* (of weedfree waterbody):
2. *Economic loss caused by the present weed infestation*:
3. *Potential loss if all available habitat was occupied by weeds:*
4. *Economic benefit from weeds, if any*:
5. *Cost of weed control by different options*:
   (a) Capital cost
   (b) Annual recurrent cost
6. *Immediate benefit of*:
   (a) Complete weed control
   (b) Any specified level of weed control
7. *Potential benefits from weed control*:

5.3: CONTROL IMPLEMENTATION

5.3.1: Selection of Time of Treatment

In general, all aquatic weeds must be destroyed before they develop reproductive propagules *e.g.* seeds, turions, and rhizomes, although in practice it is difficult to catch this state. Besides, there are relatively susceptible and tolerant stages in the ontogeny of weeds which must be considered. For example, *Nymphaea, Typha*, and *Phragmites* spp. show least regeneration when these were curbed at their active flowering stages. *E. crassipes*, on the other hand, is found most susceptible to herbicides in its active vegetative phase as occurring during the summer and monsoon periods. Submersed weeds can be exhausted more efficiently when these were mowed at young stage (Philipose, 1963). Unfortunately, scientific information on the most vulnerable stages in the ontogeny of all the different species of aquatic weeds is not available. Intensive research is needed to study responses of our major aquatic weeds to various control methods at their different physiological stages of growth.

Besides the ontogeny of a weed, the efficacy of a control measure is governed by the prevailing meteorological conditions, particularly water temperature under which a given herbicide or other control agent is to perform (Philipose, 1968). For example, the foliage-active herbicides usually work best under clear sunshine and warm day conditions. But for the soil treatment of ditchbanks, slightly moist and cool conditions prove better. Further, heavy rains received after the application of a herbicide to the ditchbank may cause run-off of the toxicant into the non-target areas. Very cold (below 10°C) or very warm (over 33°C) environment often impairs the efficacy of many a herbicide.

Period of heavy flood is largely unsuitable for aquatic weed control campaigns, except when the flood is used to drift the heavy weed mats towards their easy collection points.

Finally, a weed control operation should be so timed as to make the weedfree water available during the period when it was needed most by its users. Without a constant follow up programme, a serious weed infestation would again develop. Not only this, upon initial clearance of a waterbody there may be invasion by some new weed species. It is the law of nature that any niche in an ecosystem is tapped by some organism. And the weedy plants are most perfect in doing so, both the terrestrial and aquatic species. As such, we must be able to plan for a decreasingly intensive sequence of aquatic weed control techniques. Our efficiency lies in adopting an annual, integrated weed management programme that would hold back aquatic weeds long enough to maximise economic use of the waterbody.

5.3.2: Herbicide Efficacy Variables

Even when same herbicide is used against a weed species in different waterbodies, the results may vary markedly. It is so because herbicide effects are very sensitive to the stage of the weed growth, limnology of water, co-solvents and wetting agents present in the formulation, and nature of coverage of the weed with the herbicide. All this calls for standardisation of herbicide application techniques on aquatic weeds and study of several variables affecting the herbicide performance.

5.4: LONG TERM MANAGEMENT

No aquatic weed control technique, including the use of a soil sterilant, will bring about permanent freedom from weeds, although some techniques may postpone their regrowth more than the others. Long term control of aquatic weeds requires heavy initial clearance, followed by periodic maintenance efforts to remove the regrowth.

The management of aquatic weeds over long periods should include preventive, smothering, softening, and direct control techniques in appropriate proportions and in right sequences, depending upon several weed and ecological factors. This requires a good knowledge of the entire ecosystem in and around a waterbody. Future aquatic weed scientists must study factors that promote the growth of aquatic plants and affect the hydrobiological conditions in greater details in order to develop an integrated approach to the aquatic weed management. The thinking that modern herbicides alone could solve our aquatic weed problems is not correct. These can at best provide temporary relief to the waterbodies. The sad story of *E. crassipes* control in USA with herbicides amply supports this view. In Florida alone 10-15 million dollars were spent annually to control this weed. Still Florida had about 40,000 ha of her water covered with waterhyacinth before the scientists took to its biological control (Frank, 1976).

There is a growing evidence that to minimise the potential aquatic weed problems, a **judicial integration** of various techniques, including manipulation of plant environment, need be instituted. Many good examples of success with combinations of methods can be cited. For instance, in the labour-intensive countries, manual lifting of waterhyacinth (*E. crassipes*) mats can be made easy by pre-treating their shoots with low doses of 2,4-D to soften them. A combination of anhydrous ammonia and manual labour works the same way against mixed growth of a variety of submersed weeds.

Herbicides and growth substances have been used in specific cases to augment biological control of aquatic weeds (Center *et al.* 1982). Growth substances are also found to improve the efficacy of certain herbicides *e.g.* of 2,4-D against waterhyacinth (Pieterse and Roorda, 1982). Also, often a combination of two or more herbicides have proved superior to a straight herbicide on the target vegetation. A very good example of this approach is the successful control of hydrilla with a combination of diquat and copper sulphate. On ditchbanks, herbicide combinations are used more frequently than straight herbicides. Different physical methods can likewise be combined with herbicides and bioagents to achieve more efficient aquatic weed control than is possible with any one tool.

Further, each waterbody is typical in its environment and biotic components. Therefore, while general principles and methods of aquatic weed management can be used for the management of water resources to meet specific objectives, these may require modification in the light of local conditions of a waterbody. Therefore, any aquatic weed management programme against even similar weed problems must be planned with provision of adequate flexibility and contingency adjustments.

5.5: ECOLOGICAL SHIFTS

The phenomenon of ecological shifts in an aquatic environment is recognised as sudden growth of one or more new species of plants upon destruction of the existing ones. Several instances of shifts in weedy aquatic flora are on record which compel scientists to search for new weed control possibilities. The phenomenon is most commonly noticed when same herbicide is repeatedly employed to suppress one or more, specific weed species. Never-the-less, even physical and biological methods are prone to this phenomenon. For instance, it is commonly observed that a mat of floating weeds covering large area of water surface usually does not permit the growth of submersed weeds. But as soon as the cover of this floating weed is sunk or removed by any method there is every chance of an ecological shift towards the development of submersed plants. Removal of the hyacinth mats has been observed to also encourage the lemnid and azolla mats, which pose different kinds of problems. Mowing the submersed or ditchbank weeds likewise will tilt the balance slowly in favour of the weed species capable of withstanding frequent cuttings, but it will go against those species which succumb to it. For such reasons, aquatic weed management strategies have to be made on long-term bases, instead of feeling satisfied with only eliminating the current aquatic weed species.

SELECTED BIBLIOGRAPHY

22. Aboaba, F.O. 1973. Management of aquatic vegetation. World Crops 25 (1): 21-24.

23. Center *et al.* 1982. Control of waterhyacinth with *Neochetina eichhorniae* and growth retardant. Weed Sci. 30 (5): 453-457.

24. Frank, P.A. 1976. Distribution and utilisation research on tropical and subtropical aquatic weeds in United States. *IN* Aquatic Weeds in S.E. Asia. Dr. W. Junk b.v. Publishers, The Hague, 353-360 pp.

25. Johannes, H. 1972. Problems in the control of aquatic plants. Weed Abst. Vol. 22, Item 2057.

26. Mitchell, D.S. 1979. Assessment of aquatic weed problems. J. Aquat. Plant Manage. 17: 19-21.

27. Mitchell, D.S. 1979 (a). Formulating aquatic weed management programmes. J. Aquat. Plant Manage. 17: 22-24.

28. Philipose, M.T. 1963. Making fish ponds weedfree. Ind. Livestock 2: 20-21 and 34-36.

29. Philipose, M.T. 1968. Present trends in the control of weeds in fish culture waters in Asia and the Far East. F.A.O. Fish Rep. 44 (5): 26-52.

30. Pieterse, A.H. and F.A. Roorda. 1982. PGR sprays boost 2,4-D activity against waterhyacinth. Aquat. Bot. 13 (1): 69-72.

31. Robers, H.A. (Ed.). 1982. Weed Control Handbook: Principles. The control of aquatic weeds. Oxford, U.K., 438-448 pp.

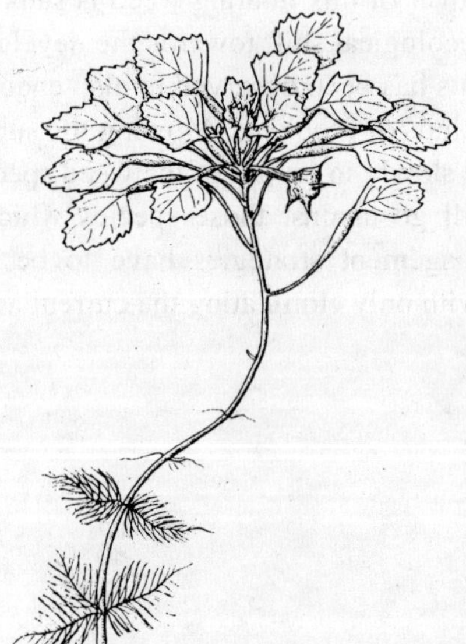

Trapa natans (Water chestnut)

Chapter 6

PHYSICAL AND UTILISATION CONTROL OF EXCESSIVE AQUATIC VEGETATION

6.1: PHYSICAL CONTROL

Physical control of aquatic weeds is generally regarded as their cutting or dislodging, followed by their removal from the waterbody. The technique also includes suitable barrier restraints to check drifts of floating aquatic plants, and other similar devices.

Manual methods like hand-pulling and hand-cutting of aquatic weeds were man's earliest efforts towards their control. Today, a number of specialised equipments for crushing, dredging, mowing, harvesting, and lifting of aquatic weeds are available for their mechanical control in different situations. These equipments take dredgery out of human life and provide more efficient weed control. The following pages briefly describe (i) the physical methods of aquatic weed control presently in vogue and (ii) the possible opportunities for utilisation of the harvested vegetation.

6.1.1: Advantages And Limitations Of Physical Control

There are several distinct advantages and limitations of physical control of aquatic weeds as follows:-

(a). Advantages

1. Physical control usually does not contribute to environmental pollution or damage to the non-target organisms.

2. Physical methods yield immediate results, without any time lag.

3. Physical methods are largely non-selective, with fewer chances of permitting ecological shifts in aquatic flora.

4. Physical removal of aquatic weeds lessens the massive nutrient load of eutrophic waterbodies, helping thus indirectly gradual reduction in future weed populations.

5. Physical control reduces dependence on costly herbicides.

6. Many times the harvested aquatic weeds can be utilised as animal feed, manure, and energy source to pay for the cost of harvesting them (see Sec. 6.2).

7. Physical methods can be practised in any localised area of a waterbody, including the flowing canals and other streams.

(b). Limitations

1. Physical removal of aquatic weeds is a labour-intensive method, suited only to countries with plenty of cheap labour at command.

2. Physical control of aquatic weeds is like mowing a lawn; the weeds regrow fast. Even digging the exposed bottoms of waterbodies is made ineffective by rapid regrowth of weeds from rootstocks, rhizomes, and tubers.

3. Physical removal of aquatic weeds, especially with machines, often induces their dissemination to new sites.

4. Repeated physical removal of weeds from a waterbody may eventually deplete it of its plant nutrients and, thus, reduce the growth of phytoplanktons. However, in practice this is not a serious limitation since one can always resort to fertilisation of the waterbody.

6.1.2: Physical Control Techniques And Tools

(a). ***Chaining***: In this method the ends of a heavy drag chain, cable, A-shaped frame drag, or chain scythe are tied to either two powerful boats or bulldozers, depending upon the width of the waterbody. The drag chain (or drag frame) is then pulled close to waterbottoms in up and down trips, breaking thus the rooted submersed weeds into fragments. The crushed weeds float to the surface of water from where these are collected manually or mechanically. In the case of small waterbodies, short chain arrangements can be managed by two men. Chaining is effective in the destruction of long, slender, submersed aquatic weeds.

(b). ***Dredging***: Dredging is the pulling of weeds alongwith their roots from the mud. Mechanical clean-out buckets and hand-operated weeding forks and showels are used for this purpose. Dredging is a slow, untidy, and more expensive method than chaining, but it removes the weeds more completely from ditches and ponds. The dredging equipment can also be used for picking the floating weed mats.

Of recent, a machine called 'weed witch' has been developed that combines the effect of chaining and dredging of rooted aquatic weeds. It dislodges the weeds from the hydrosoil under highly forceful jets of water released from the high speed centrifugal pumps. Gasoline engines supply the needed power to the pumps.

(c). ***Draining***: It is a common practice in India to drain the weedy canals during the summer months to expose their bottoms and collect the weeds manually, employing thousands of labour in the operation. The dry canal bottoms are then ploughed in the hope of desiccating the weed propagules under hot sun. However, the method, besides being expensive, is not sufficiently effective. The weed propagules never get completely killed and by monsoon season the weeds grow back to their original levels.

(d). ***Netting***: Small, floating weeds like duckweeds and waterfern can be skimmed out of small waterbodies using **drag-nets.** For manual operation from boats the nets are usually made of 3 mesh coir ropes.

(e). ***Barriers***: Fencing with materials like bamboos and inflatable rubber booms are used to restrain the drift of a free-floating aquatic weed *e.g.* waterhyacinth. The barriers are so made as to permit water to pass through them and to conform to wave and wind actions. The weed barrier can also be created with a rope threaded through cork floats, or with a chain of barrels.

(f). ***Water weed cutters and harvesters***: In high discharge canals and very large lakes, physical control of submersed and floating weeds has been made easy and economical by the development of a variety of underwater weed cutters and harvesters in certain agriculturally advanced countries. In fact, in ditches of over 2.8 cumsec flow capacity the application of herbicides usually proves ineffective and here the use of weed cutters and harvesters is the major choice for controlling the rooted, submersed vegetation.

(i) *Underwater weed cutters*: An **underwater weed cutter** is a motorboat mounted equipment with a sharp, strong cutter-bar, having reciprocating blades sliding against a fixed blade. The cutter-bar is heavily powered and it cuts weeds clean, about 1 m below the water surface (Plate. 6.1). There are several variations to the cutter-bars with competitive advantages and disadvantages. These are sold in different countries under trade names like `Water Bug', `Sawfish', `Boat Sweeper', `Wilder Weed Cutter', etc. The harvested weeds float to the water surface which are then easy to collect. Manually operated underwater weed cutters are also available. The underwater weed cutters work efficiently when (i) these are not obstructed frequently by bridges and like structures, (ii) the velocity of water is no more than 60 cumsec, and (iii) the weedy waterbody is only a little deeper than the vertical reach of the cutter-bar of the weed cutter. If the depth of a weedy waterbody is 3 m and the cutter-bar works only 1 m deep, obviously considerable portions of submersed weeds would remain unsevered. In Chambal canal in Rajasthan (India), Wilder Weed Cutter was found to clear one ha water of its submersed weeds in about 10 hours. Also, the use of the machine was found more economical than the conventional method of draining the canals and digging their bottoms. Moreover, the technique was attractive since it kept the canals going round the year. In these canals, integration of weed cutters with biocontrol of weeds with grasscarp (Chapter 8) has been found a very effective scheme to keep their water flowing freely. In West Bengal, weed cutting launches having V-shaped sickles have been used for cutting *Colocasia antiquorum* and other tall weeds in shallow waters. Power winches are employed in Orissa to eradicate dense growth of rooted, submersed vegetation.

(ii) *Harvesters*: Any machine that either cuts or picks up aquatic weeds from a waterbody and conveys these to the shore simultaneously, is called a **harvester.** It can be either mobile or shore-based. It is usually powered with a 15-30 H.P. machine. A mobile harvester is launched on a sturdy boat to pick-up weeds and to convey them to the shore. It is usually a low capacity weeder. A common trade mark of mobile harvester designed specifically for removing waterhyacinth is `Hybailer'. Several others are marketed in the western world.

Plate 6.1: An under-water weed cutter in operation in a canal at Kota (Photo-author)

Plate 6.2: A weed cutting bucket showing reciprocating cutter bar on its leading edge. (Courtesy: T.O. Robson, ADAS, USA)

Plate 6.3: A highly mechanized, off-shore hyacinth harvester in operation somewhere in USA. It simultaneously lifts the hyacinth mat, breaks it, and throws the disintegrated plants on shore (Courtesy; NAS, U.S.A.). Several other modifications are available and used in highly mechanized countries for clearing the waterbodies of their hyacinth infestations

The shore-based harvesters are stationed at convenient pick-up points down the water current. The floating mats of a weed like waterhyacinth are pushed to this location with the help of drift boats equipped with cables and ropes. Sometimes, water currents in a waterbody can be utilised as aid to push the weed mats down the stream. The harvester picks-up the drift-fed masses of the weed and heaps them on the shore. The shore-based harvesters are high capacity machines. Many times the accumulation of weeds on the shore may be so fast as to cause shortage of space there. To avoid this inconvenience certain harvesters are further provided with flat-bed trucks to carry away the weeds from the site of operation. One such equipment is `Aqua-Trio'. It is a very efficient, but expensive, system. Besides avoiding accumulation of the weed heaps on the shore, it minimises chances of weed dispersal around the waterbody.

Some harvesters are combined with a bailing equipment that sucks-in the floating weeds, presses them, and then turns them into bails of convenient sizes for their easy transport. Such harvesters are complex and costly machines. The agricultural engineers in the developing nations can render great service by developing low cost, easy to operate, and simple to maintain aquatic weed harvesters.

6.2: UTILISATION CONTROL

The utilisation of aquatic weeds is, in fact, an indirect method of their control and it should be given serious consideration. Government incentives to develop and practice new methods of utilisation of aquatic weeds will be greatly helpful in controlling aquatic weeds the non-pollution way and restoring our waterbodies to their intended uses.

In parts of the tropics and subtropics summer vegetation is limited on land but it is abundant in water. A complete mat of waterhyacinth, for example, has been found to produce 350-1700 t ha^{-1} green vegetation, growing at a remarkable rate of 5 q ha^{-1} day^{-1} (Westlake, 1963). Submersed weeds have an average annual productivity potential of 25-350 t ha^{-1} of green matter (Table 6.1). An infestation of cattails (*Typha* spp.) can yield up to 500 t ha-1 of vegetation (Steward, 1970). Obviously, it will be a mistake to let these enormous quantities of vegetation go waste, besides being bothered by them.

Table 6.1: Biomass Production Potential of Some Aquatic Weeds (Zutshi and Vass, 1971; Steward, 1970; Kaul and Bakaya, 1976; Matai, 1976)

Weed species	Annual productivity $t\ ha^{-1}$ (dry weight)
Allmania nodiflora	3.3
Ceratophyllum spp.	3.8
Cyperus papyrus	83.4
Eichhornia crassipes	10-44
Hydrilla verticillata	24-50
Ipomoea reptans	8.0
Lemna spp.	1.0
Myrlophyllum spp.	2.0-2.5
Phragmites communis	33.5
Pistia stratiotes	4.5
Sagittaria subuleta	23
Salvinia spp.	3.1-6.5
Thalassia testudinum	33
Typha latifolia	50.6

In view of the huge biomass production potential of aquatic weeds, there could be many suggesions for their commercial utilisation but all future projects aimed at utilising aquatic weeds for controlling them must fulfil the following five pre-requisites. Besides, any approved project on utilisation of aquatic weeds should be initially subsidised by the local governments.

1. *The technology should be such as would require extensive consumption of the aquatic weeds; fast enough to outstrip their self propagation.*

2. *The technology should be economical and remunerative enough to at least partly defray the cost of physical removal of aquatic weed/s.*

3. *The proposed processing of aquatic weeds should be independent of seasonal variations.*

4. *The proposed on-shore processing plant should be easy to shift to new sites when the population of the concerned weed has gone low at the initial site.*

5. *The proposed utilisation of aquatic weeds should not lead to their dispersal to new sites.*

Keeping the above five riders in view some promising options for commercial utilisation of aquatic weeds are as under:-

6.2.1: Aquatic Weeds As Compost And Soil Conditioners

Several aquatic plants can form valuable source of plant nutrient-rich compost (Table 6.2). But, thus far, waterhyacinth alone has been explored for the purpose in agriculture.

Table 6.2: N and P Contents and per ha potential of Certain Aquatic Weeds on Dry Weight Basis (After Steward, 1970)

Weed species	N (%)	P (%)	N (kg ha^{-1})	P (kg ha^{-1})
Cyperus papyrus	1.5	0.15	1250	125
Eichhornia crassipes	4.0	0.4	1400	140
Hydrilla verticillata	4.0	0.4	100	10
Myriophyllum exalbescens	3.9	0.4	98	10
Phragmites communis	1.5	0.15	490	50
Sagittaria subulata	4.0	0.4	916	91
Thalassia testudinum	4.0	0.4	1324	132
Typha latifolia	1.5	0.18	760	90

In general, the average N, P, and K contents of aquatic weeds are accepted as 1.5-4.0% N, 0.2-2.0% P, and 0.15-4.9% K (Mazid, 1983). Likewise the ranges of these nutrients in algae are found as 1.3-4.9% N, 0.08-0.17% P, and 0.06-0.69% K. Considering these plant nutrient contents of aquatic weeds, turning them into compost naturally seemed a priority item in their utilisation options. A start was made with waterhyacinth (*E. crassipes*) because of its extensive availability in the country. Its plants have C, N ratio of 20:1 to 30:1 and contain about 3.5% organic matter, 0.04% N, 0.06% P_2O_5, and 0.02% K_2O. India has potential of 3.0 million tonnes hyacinth compost, annually. In comparison to F.Y.M., waterhyacinth compost is richer in potash. The main hurdle in composting waterhyacinth, as also other aquatic weeds, is its high water content of over 90%. This leads to anaerobic decomposition of the material in piles. To activate aerobic decomposition of waterhyacinth, frequent stirring of the piles is essential which is a labour intensive process. To improve upon the situation, it is advised that before preparing the compost piles, waterhyacinth should be left on shore for a couple of days to wilt. Then, during piling of the wilted hyacinth, a suitable inoculum like well rotten compost, dung, sewage sludge, or even decomposed waterhyacinth dug from bottoms of ditches, should be spread after every meter height of the pile.

In India, a simplified method of composting waterhyacinth in pits was developed by the Assam Agricultural University which avoided frequent stirring of the piles. In this method the pits are dug on elevated but levelled ground. These pits taper towards the ground and are usually 3 m x 2 m at the bottom and 15 m deep. Each such pit can accommodate about 8 t of raw hyacinth. The hyacinth plants are chopped into about 15 cm long pieces and left to wilt for sometime. Long bamboos are erected in pits at convenient distances and the wilted hyacinth plants are filled in the pits. After 1 m filling of the pits, the piles are inoculated with dung or like material. Periodically water is sprinkled on the piles. After 10 days or so, additional wilted hyacinth plants are added in the pits to fill the spaces made by the sinking of initial material. Then the pits are packed with soil and the bamboos are slowly pulled out to leave back enough aeration holes. The compost should be ready in about 3-4 months.

There is another method of composting waterhyacinth which is known as the 'Indore Method'. In this method the aerobic decomposition is accomplished by reducing the depth of the pit to under 2 m and that of each layer of waterhyacinth to 30 cm. No bamboo holes are made in the pit in this method. The duration of composting in the Indore method is 3-5 months.

In some countries, instead of making the dug-pits, bamboo cages of almost the same dimensions as the pits are erected on the ground to contain the layers of waterhyacinth and dung. This avoids the manual diging of the pits in the ground.

Field research in Ceylon has shown that the waterhyacinth compost is effective in improving the crop yields and earthworm population in the treated soil (Balsubramanian and Bai, 1995 and Mazid, 1992). In India it is reported promising in the reclamation of saline-alkali soils, in conjunction with sludge and pyrite (40 GR).

In USA, aquatic weeds, particularly waterhyacinth, have been turned into easy to handle soil conditioners by their energy-intensive, mechanical drying and grinding. Considerable expenditure is incurred on the transport and processing of the aquatic weeds.

Presently, it seems that conventional composting (or vermi-composting) is a very promising avenue of utilising aquatic weeds but, somehow, it has not caught up so widely in the country as it should have. This calls for sincere Research and Development (R and D) efforts on the part of soil scientists in India to find the shortfalls and remedies for its large scale adoption, encouraging thus indirectly the physical control of weeds in waterbodies and simultaneous improvement of our agricultural lands. There is one caution here. The final compost product should be devoid of any living aquatic weed propagule, particularly if the compost was to be used on constantly moist lands, like in paddy and jute cultivation, for obvious reasons.

Further, to hasten the process of composting of waterhyacinth, and other aquatic weeds, microbiologists can help in isolating the microorganisms concerned in the process and prepare their concentrated inoculums. Similarly, the agricultural engineers can develop small, easy to maintain, machines to remove aquatic weeds from waterbodies and dewater them.

In Kanya Kumari (Tamil Nadu) waterhyacinth is successfully used year after year as green manure in standing banana plantations on large scale. Interestingly enough, the waterhyacinth plants are brought for this purpose from neighbouring districts.

6.2.2: Aquatic Weeds As Herbivorous Fish Food

In the last 2-3 decades, Chineese grasscarp (= amur) has shown great promise in mass consumption of several species of aquatic plants (chapter-8) and turning these into meat protein for man. While this method of utilisation of aquatic weeds, particularly of the submersed types, has been exploited in many countries, in India, on some pretext or the other, it has not been given adequate importance and action it deserves. Specific R and D projects comprising fishery experts should be developed on the utilisation of aquatic weeds for fish meat production. Main emphasis has to be given on the standardisation of artificial multiplication of the grasscarp and determination of optimum conditions for its release in different kinds of waterbodies.

6.2.3: Aquatic Weeds As Animal Feed

On dry matter basis, aquatic weeds are found to contain 10-26% crude protein (C.P.), although some of these like *E. crassipes* and *Lemna* spp. may contain even 25-35% C.P. in their leaves. The proteins in aquatic weeds are usually low in their lysine and methionine contents. *Hydrilla* and *Myriophyllum* are additionally rich in carotene and xanthophyll. Some scattered records on nutritional constituents of aquatic weeds are compiled in Tables 6.3 through 6.5.

The three main bottle-necks in utilising aquatic weeds as processed feed are their high moisture contents, high mineral contents, and the presence of toxic elements in some of the species.

(a) ***High moisture content***: Fresh samples of most of the aquatic weeds contain as much as 95% water. This makes them cumbersome to transport and impossible to store without mould. Therefore, aquatic weeds must be dewatered to about 30% moisture before processing them into animal feed on a commercial scale. In semi-arid areas, solar drying of aquatic plants may be feasible by exposing them to sun in thin layers but this is a very slow process. In the U.S.A., mechanical dewatering of aquatic weeds by macerating and pressing machines has been devised. It is sometimes more useful to combine solar energy with machines to achieve the 30% moisture level.

Some aquatic weeds may be found suitable for making silage. In that case their intensive dewatering is not required. For example, good silage has been made out of *E. crassipes* at 50-90% moisture level (see Chapter-11).

Table 6.3: Dry Matter and Protein Contents of Certain Aquatic Weeds (After Boyd and Blackburn, 1970; Little and Henson, 1967; Pirie, 1970; Taylor *et al.*, 1971)

Weed species	Dry matter (%)	Crude Protein (% d.w.)
Alternanthera philoxeriodes	8.2-15.0	8.1-12.8
Ceratophyllum demursum	-	16.9-18.8
Eichhornia crassipes	-	-
(i) whole plant:	4.0-9.3	4.7-23.5
(ii) leaves:	-	31.3
Elodea canadensis	5.8-12.0	13.2-39.4
Ipomoea reptans	-	28.2-36.3
Justicia americana	-	10.0-23.5
Lemna minor	-	15.6-31.3
Myriophyllum spicatum	-	11.3-25.6
Najas guadelupensis	8.8-14.4	17.4-41.6
Nuphar advena	7.5-12.4	9.3-13.3
Phragmites communis	-	11.3
Pistia stratiotes	5.8-8.3	10.6-24.7
Potamogeton spp.	-	6.9-21.9
Salvinia sp.	7.2-10.7	5.0-11.3
Typha sp.	-	3.1-15.0

***(b) High mineral content*:** Aquatic weeds are very high in their mineral (= ash) contents. Most of the submersed weeds contain 60% ash. Waterhyacinth has an ash content of little over 20%. The ash of aquatic weeds is usually 10 to 100 times richer in sodium; 4 to 19 times higher in iron; and 3 to 6 times so in potassium than many food crops. Kimbal and Bakar (1983) reported that on dry weight basis Na and K contents of *Myriophyllum heterophyllum* were 0.2-2.7% and 1.4-3.3%, respectively.

The high ash contents in aquatic weeds limits their utilisation as sole animal feed. Weeds can at best supplement 10-12% of total daily feed of animals. This is also true of waterhyacinth, which has been the most common aquatic plant under research with the animal nutritionists. However, waterhyacinth from clean, nutrient-rich waters is found to be comparatively low in its ash content whence it can form up to 25% of daily animal feed.

Table 6.4: Composition of *Eichhornia crassipes* (whole plant) (Compiled from Frank, 1976)

% Fresh weight		% Oven dry weight		% Processed-dry	
Moisture	= 92-95.6	Carbon	= 35	Organic matter	= 89
		Crude protein	= 16		
Dry matter	= 4.5	Ash (inorganic matter)	= 25	Ash	= 11
		Cellulose	= 28		
Organic matter	= 3.5	N (Available)	= 1.6%	Crude protein	= 10
		P	= 0.3%		
		K	= 3.8		
Inorganic matter (Ash)	= 1.0	Na & Mg	= 0.6		
		C:N	= 23:1		

Table 6.5: Composition of *Hydrilla verticillata* (Compiled from Frank, 1976)

Unprocessed, oven-dry plants (%)		Processed, air-dry plants (%)		Fresh plants (%)	
Crude protein	= 18	Dry matter	= 72	Moisture	= 92
Cellulose	= 33	Ash	= 28	Dry matter	= 8
Available CHO	= 7	Crude protein	= 13		
Ash	= 27				
Ether extract	= 3.5				
P	= 0.3				
Ca	= 4.4				
Mg	= 0.9				
K	= 2.9				

***(c) Toxic elements in weeds*:** Not all aquatic plants can be used as animal feed indiscriminately. Many plants may prove fatal to animals for one reason or the other. Even the environment in which the weeds grow, or the stage at which these are harvested, can determine their toxicity potentials. Some weeds are known to accumulate up to 1000 ppm or more of arsenic in their tissues which makes them unfit as animal feed. In certain weeds, and at certain times, nitrates may accumulate to toxic levels. For instance, Tucker and Debusk (1983) reported that NO_3 content of *E. crassipes* ranged from 0.05-0.21% (d.w.b.) in different seasons and the highest NO_3 levels were reached during winter when its growth rate was the lowest. At that time it was unsafe to feed this plant to animals. Besides, there seems to be some other incriminating factor in waterhyacinth which limits its use as a sole animal feed. Even pigs fed entirely with this plant may develop diarrhoea.

Aquatic weeds growing in polluted waters are usually more harmful to animals than those growing in clean waters.

The use of waterhyacinth as regular forage for ruminants has been investigated in considerable depth, particularly in India, Ceylon, Sudan, and Bangladesh. Now it is finally accepted that in its raw form it cannot replace, even partially, the conventional feed of the ruminants, largely due to its high content of alkali metals. Some scientists have found that prewilting of waterhyacinth to about 30% moisture level, followed by addition of rice straw, mustard cake, and sugarcane molasses can make it suitable to supplement the conventional animal feed of cattle and bufalloes to the extent of 20-25%. Waterhyacinth can also be turned into silage by the addition of sugarcane molasses or rice bran (Lowilai *et al.*, 1995). Alligatorweed, likewise, has been found suitable to replace the conventional animal feed up to 15% with such treatment (see also Chapter-11).

It is obvious that lot more research is needed to utilise aquatic weeds as animal feed than those needed in using these for composting. The on-shore dewatering and blending machines are needed to blend weeds with suitable additives and turn these into easy to handle, marketable forms. Besides, new additives are to be discovered to improve the proportion of voluntary intake of the prepared feed. Specific treatments may also have to be developed to remove the alkali metals and other toxic substances from aquatic weeds so that the final feed can be rated as acceptable *in terms of the actual milk and meat output of the animals fed upon it.*

6.2.4: Aquatic Weeds As Source of Paper Pulp

The use of aquatic weeds for making pulp for paper and insulation boards at industrial level has been considered. For this purpose we need aquatic weeds with strong, hydrophobic fibre. Waterhyacinth is unsuitable for paper making because it is a low fibre yielder (13%) and its fibre is hydrophilic. Also, the paper made out of it wrinkles and darkens on drying. It is brittle and possesses poor tear properties (Nolan and Kirmse, 1974; Frank, 1976). Other aquatic weeds should be studied for their paper, cardboard, and rayon pulp qualities.

In Romania, common reed (*Phragmites communis*) is used on large scale to make printing paper and cardboards. The by-products available from the reed paper industry include cemented reed blocks, compressed fibreboards, and insulation materials. Over one lakh tonnes of reed is used annually in *Rommania* for the purpose. The air-dry reed yields about 60% pulp because of its high cellulose content in leaves (about 50%). It is considered suitable also as source of rayon pulp. Long back in USA too considerable paper pulp was made out of *Typha* spp. (cattails). The cattail pulp makes good, strong wrapping paper. Its fibre is long like that of jute and it constitutes 30-40% of the plant on air-dry basis.

Cyperus papyrus and *C. antiquorum* are other aquatic plants that were used in ancient times for making paper. Their large pith, although difficult to separate, can be used for making hardboards. Attempts are underway to use *Cladium mariscus* and *Scirpus* spp. as sources of paper pulp. In Kerala (India), *Salvinia molesta* (water fern) has been found to form good raw material for making paper, cardboard, fibreboard, and road stabilisers.

6.2.5: Aquatic Weeds As Source Of Energy

Scientists have been successful in using waterhyacinth as source of methane-rich biogas. The idea was seeded from natural production of methane, seen as bubbles, in marshes where dead weeds decompose under anaerobic conditions. The chief attraction in the proposal is the utilisation of aquatic weeds without having to dewater them. In fact, the high moisture content of aquatic weeds is helpful in maintaining their anaerobic, bio-decomposition in the chambers to produce methane. (Under aerobic conditions, instead of CH_3, CO_2 is formed). In the pioneering work conducted on bio-conversion of waterhyacinth into methane in USA it was estimated that each kg of waterhyacinth on dry weight basis yielded about 370 l of biogas, with 69% methane (Casfbow, 1967).

Although a large number of biogas plants are in operation in India, China, and elsewhere, these can process mainly cattle dung. Waterhyacinth is observed to choke their inlet and outlet pipes, besides inducing algal growths. The Central Mechanical Engineering Research Institute at Durgapur (W. Bengal) has constructed a biogas plant especially suited to process waterhyacinth but it needs cleaning twice a year. Further work to develop a waterhyacinth biogas plant is in progress there.

Further, it is now established in India (and China) that to obtain high quality biogas it is required to mix chopped and dried waterhyacinth with cattle dung or urine in ratios ranging from 1:1 to 3:2. It is also claimed that biogas liberation from the mixture is less dependent on temperature changes than gas production from cattle dung alone. China has made remarkable progress in the field of waterhyacinth biogas equipments. Their important features are: (a) combined gas holders and digesters and (b) large inlet and outlet openings.

A major difficulty in installing the biogas plant near any waterhyacinth-infested waterbody is that once the waterbody is cleared of this weed, the equipment has to be shifted to a new site. Besides, the gas production from the plant will be much limited during the winter season. Probably, because of such reasons this method of utilisation of aquatic weeds has not caught momentum, thus far.

6.2.6: Aquatic Weeds As Source Of Proteins And Carotene

The crude protein (C.P.) contents of many aquatic weeds are high. In general, Mazid (1983) reported macrophytes like *Eichhornia, Pistia*, and *Salvinia* spp. contained 10-25% of C.P. while duckweeds showed 15-28% of it. Taking into consideration their biomass production, macrophytes had potential of producing up to 650 kg C.P. and duckweeds up to 200 kg C.P. per ha per month. Certain green algae can produce even up to 1166 kg C.P. per ha per month. Such data induced the scientists to develop methods to extract proteins from aquatic weeds (Pirie, 1970 and 1973; Taylor *et al,* 1971, and Matai, 1976). The protein extractability varied from 6.3-38% of the original content. The extraction was usually difficult from the whole plants but it was comparatively easy from their leaves.

Besides developing economical methods of extracting proteins from aquatic plants, determination of their actual digestibility by animals and man in terms of `digestible proteins' will require considerable research. May be, similar efforts are needed for the extraction and utilisation of carotenes from aquatic weeds.

SELECTED BIBLIOGRAPHY

1. Ali, M.A. and S. Leeson. 1994. Nutritional value and utilisation of aquatic weeds in the diet of poultry. World Poultry Science J. 50 (3): 237-251.

2. Anjana, D. and M. Shyam. 1996. Nutritional evaluation of leaf protein extracted from three aquatic plants. J. Agr. and Food Chem. 44 (8): 2162-2166.

3. Anomyous. 1978. Making Aquatic Weeds Useful--Some Prospects for Developing Countries. National Academy of Sciences, Washington D.C., USA, 175 p.

4. Bagnall, L.O. *et al.* 1974. Processing and storage of waterhyacinth silage. Hyacinth Control J. 12: 73-79.

5. Balasubramaian, P.R. and R.K. Bai, 1995. Recycling of cattle dung from biogas plant and waterhyacinth in vermiculture. Bioresource Technol. 52 (1): 85-87.

6. Baldwin, J.A. *et al.* 1974. Preservation and cattle acceptability of waterhyacinth silage. Hyacinth Control J. 12: 79-81.

7. Bhanthumnavin, K. and M.G. Mc-Garry. 1971. *Wolffia arrhiza* as a possible source of an expensive protein. Nature 232 (5311): 495.

8. Bock. J.H. 1969. Productivity of waterhyacinth. Ecology 50: 460-464.
9. Boston, H.L. and M.A. Perkins. 1982. Water column impacts of macrophyte decomposition beneath fibre glass screens. Aquatic Bot. 14 (1): 15-27.
10. Boyd, C.E. 1968. Freshwater plants—a potential source of protein. Econ. Bot. 22 (4): 359-368.
11. Boyd, C.E. 1969. The nutritive value of three species of water weeds. Econ. Bot. 23 (3): 123-127.
12. Boyd. C.E. and R.D. Blackburn. 1970. Seasonal changes in the proximate composition of some common aquatic weeds. Hyacinth Control J. 8 (2): 42-44.
13. Bryant, C.B. 1973. Control of aquatic weeds by mechanical harvesting. PANS 19: 601-605.
14. Casfbow, A.J. 1967. Convert waterhyacinth to methane PANS (C) 13 (3): 243-244.
15. Chibbar, S.S. and G.D. Singh. 1971. Paddy straw and waterhyacinth silage. Ind. Fmg. 20 (11): 24-26.
16. Days, M.N.G. and N.W. Pirie. 1960. Protein from the leaves by bulk extraction. Engineering 190, 274-275.
17. Deutsch, A. 1974. Some equipment for mechanical control of aquatic weeds. International Plant Protection Centre (IPPC), USA, Rept. No. 74-2, 17 pp.
18. Frank, P.A. 1976. Distribution and utilisation research on tropical and subtropical aquatic weeds in U.S. (*IN*) Aquatic Weeds in S.E. Asia, Dr. W. Junk b.v. Publishers, The Hague 353-360 pp.
19. Garg, P. *et al.* 1994. Chromium-induced changes in *Limnathemum*-a possible bioindicator. Phyto-morphology 44 (3-4) 201-206.
20. Gaur. S. *et al.* 1991. Relative contributions of bacteria and fungi to waterhyacinth decomposition. Aquatic Bot. 43: 1-15.
21. Ghosh, J.J. 1967. Leaf protein concentrates: problems and prospects in the control of protein malnutrition. Trans. Bose Res. Instt. 30: 215-228.
22. Goswami, T. and C.N. Saika. 1994. Waterhyacinth--potential source of raw material for grease-proof paper. Bioresource Technology 50 (3): 235-238.
23. Hanisak. N.D. *et al.* 1980. Recycling of the nutrients in residues from methane digesters of aquatic macrophytes for new biomass production. Resource, Recovery, and Conservation 4: 313-323.
24. Kamal, T.A. and E.C.S. Little. 1970. The potential utilisation of waterhyacinth for horticulture in Sudan. PANS 16 (3): 438-496.

25. Karim, A. 1984. Microbial decomposition of waterhyacinth. Soil Sci. 66: 401-416.

26. Kaul V. and U. Bakaya. 1976. The noxious, floating *Lemnid-Salvinia* aquatic weed complex. (*IN*) Aquatic Weeds in S.E. Asia, Dr W.Junk b.v. Pub., The Hague 183-192 pp.

27. Kimbal, K.D. and A.L. Baker. 1983. Temporal and morphological factors related to mineral composition in *Myriophyllum heterophyllum* Michx. Aquat Bot. 16 (2): 189-205.

28. Koegal, R.G. *et al.* 1973. Increasing the efficiency of aquatic plant management through processing. Hyacinth Control J. 11 (June): 24-30.

29. Krishnamoorty, U. *et al.* 1995. Nutritive values of alligatorweed and paragrass. Current Research, UAS, Banglore 24 (9): 164-165.

30. Linn, J.G. *et al.* 1975. Nutritive value of dried or ensiled aquatic plants. J. Animal Sci. 41: 601-615.

31. Littlc, E.C.S. (ed) 1968. Handbook of Utilisation of Aquatic Plants. Pub FAO, PL: CP/20, 123 p. (cyclostyled)

32. Little, E.C.S. 1970. From water weeds to milk. PANS 16 (1): 198-199.

33. Little E.C.S. and I.E. Henson. 1967. The water content of some important tropical water weeds. PANS (C): 13 (3): 223-237.

34. Lowilai, P. *et al.* 1995. The feed-ing value of waterhyacinth silage-added rice bran. Grassland Sci. 41 (1): 16-20.

35. Madamwar, D. *et al.* 1990. Effect of agricultural land wastes on anaerobic digestion of waterhyacinth-cattle dung. J. Ferment. Bioengineer 70 (5): 343.

36. Madamwar, D. *et al.* 1991. Effects of various surfactants on anaerobic digestion of waterhyacinth-cattle dung. Bioresource Technol. 37: 157-160.

37. Majid, F.Z. 1983. Aquatic weeds and algae, the neglected natural resources of Bangladesh (booklet), 26 p.

38. Majid, F.Z. 1992. Aquatic Weeds--Utility and Development. Agro Bot. Pub., 96 p.

39. Majid, F.Z. *et al.* 1992. Agricultural use (as organic manure) of algae and aquatic weeds in Bangladesh. Bangladesh J. of Sci. and Indust. Research 27 (3-4): 56-61.

40. Mani, P.K. and B. Philip. 1995. Potential application of African weed (*Salvinia molesta* Mitchell) for the cultivation of oyster mushroom. Ind. J. Expt. Biology 33 (10): 806-808.

41. Mason, C.F. and R.J. Bryant. 1975. Production, nutrient content, and decomposition of *Phragmites communis* and *T. angustifolia*. J. Eco. 63: 71-95.

42. Matai, S. 1976. Protein from water weeds. *IN* Aquatic Weeds in S.E. Asia, Dr. W. Junk b.v. Pub. The Hague 369-374.

43. Nolan, W.J. and D.W. Kirmse. 1974. The paper making properties of waterhyacinth. Hyacinth Control J. 12 (May): 90-97.

44. Pandey, D.D. *et al*. 1995. Biomass, primary productivity, energy, and nutrients by *Polygonum plebeium* of lowlands of Bihar, Env. and Eco. 13 (3): 576-578.

45. Patel, V. *et al*. 1992. Effects of absorption on anaerobic digestion of waterhyacinth-cattle dung. Bioresource Technol. 40: 179-181.

46. Pirie, N.W. 1970. Weeds are all not bad. CERES, The FAO Review 4 (4): 4.

47. Pirie, N.W. 1973. Plants as sources of unconventional protein foods. The biological efficacy of protein production. J.G.W. Jones (ed)., Cambridge University Press, 101-118.

48. Riemer, D.N. and S.J. Toth. 1970. Chemical composition of five species of Nymphaeceae. Weed Sci. 18 (1): 4-6.

49. Sale, P.J.M. and R.G. Wetzel. 1983. Growth and metabolism of *Typha* in relation to cutting treatments. Aquat. Bot. 15 (3): 321-324.

50. Schertz, C. *et al*. 1983. Harvesting cattail rhizomes as an alternative feedstock for alcohol production--Modifications of potato harvester. Final Report Bio-energy Coordinating Office of the University of Minnestoa, USDE, Washington, D.C. No. DE 83009726, 21 p.

51. Sharma K.P. and P.K. Goel. 1987. Decomposition of waterhyacinth. Int. J. Ecol. Environ. Sci. 13: 13-17.

52. Sharma, A. 1971. Eradication and utilisation of waterhyacinth--A review. Curr. Sci. 40 (3): 51-55.

53. Singh, N. and L.V. Venkataraman. 1982. Status of Research on Leaf Protein and Micro-algae in India, CFTRI, Mysore (India), 86 p.

54. Singh, R.B. 1972 (a) Generating methane from organic wastes. Compost Sci. 13 (1): 20-25.

55. Singh, R.B. 1972. Building biogas plant compost. Sci. 13 (2): 12-16.

56. Singhal, P.K. *et al*. 1992. Relative contribution of different decay processes to the decomposition of *Eichhornia crassipes* (Mart.) Solms. Aquatic Bot. 42: 265-272.

57. Solander, D. 1983. Biomass and shoot production of *Carex rostrata* and *Equisetum fluviatile* in fertilised and unfertilised lakes. Aquatic Bot. 15 (4): 349-366.

58. Steward, K.K. 1970. Nutrient removal potentials of various aquatic plants. Hyacinth Contr. J. 8 (2): 34-35.

59. Stoller, E.W. and E.J. Weber. 1979. Lipid constituents of some common weed seeds. Agri. & Food Chem. 18 (3): 361-364.

60. Taylor, K.G. *et al.* 1971. Extraction of protein from waterhyacinth. Hyacinth Control J. 8 (1): 20-22.

61. Timmer, C.F. and L.W. Weldon. 1967. Evapotranspiration and pollution of water by waterhyacinth. Hyacinth Control J. 6: 34-37.

62. Toerien, D.F. *et al.* 1984. Chemical composition of *Salvinia molesta* from eastern Caprivi strip. Madoqua 14 (1): 91-93.

63. Trivedi, R.K. 1983. Waterhyacinth for pollution control, biogas, paper pulp, animal feed, and manure. Environ. and Ecol. 1: 139-143.

64. Tucker, C.S. and T.A. Debusk. 1983. Seasonal variations in the nitrate content of waterhyacinth. Aquatic Bot. 15 (4): 419-422.

65. Vanzon, J.C.J. 1984. Economic weed control with grasscarp. Trop. Pest. Management 30 (2): 179-185.

66. Velu, M. 1969. A mechanical device for eradication of submersed aquatic weeds. Fishery Technol. 6 (1): 18-25.

67. Velu, M. 1976. Development of equipment for eradication of aquatic weeds. *IN* Aquatic Weeds in S.E. Asia, Dr. W. Junk b.v. pub. The Hague, 233-240 pp.

68. Westlake, D.F. 1963. Comparisons of plant productivity. Biol. Rev. 38: 385-425.

69. Zafar, A.R. 1976. Economic significance of certain species of *Scirpus*. *IN* Aquatic Weeds in S.E. Asia, Dr. W. Junk b.v. Pub., The Hague, 387-392 pp.

70. Zutshi, D.P. and K.K. Vass. 1971. Ecology and Production of *Salvinia natans* Hoffin in Kashmir. Hydrobiologia, Netherland 38 (2): 303-320.

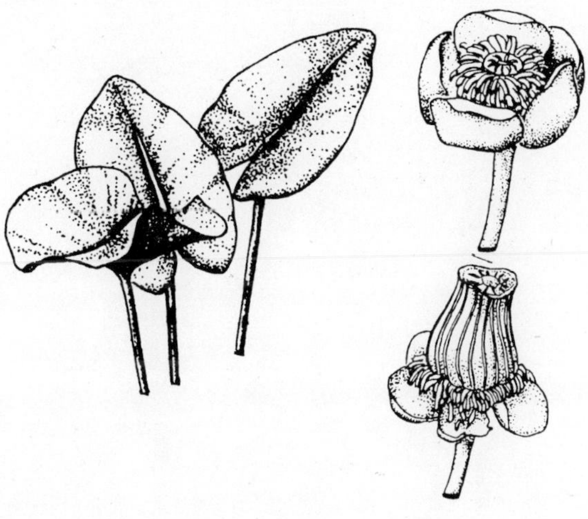

Nuphar advena (Yellow water-lily, cow lily, spatterdock)

Chapter 7

HERBICIDAL CONTROL OF EXCESSIVE AQUATIC PLANTS

In many situations herbicides can be the preferred tools for managing noxious aquatic vegetation. Their use, however, presents a number of unique problems, primarily because of the multifarious uses of our waterbodies and the possible pesticide residues in the aquatic environment. To be acceptable, an aquatic herbicide must be compatible with man's direct and long-term uses of the treated waterbody. Herbicides that control aquatic weeds and are at the same time safe to all possible uses of the treated water, are yet to be developed. Nevertheless, a suitable selection can be made from the several available herbicides, keeping in view the major weed problems, the primary uses of the waterbody, and the effect of the chemical on aquatic food down the stream.

7.1: ADVANTAGES AND LIMITATIONS OF AQUATIC HERBICIDES

There are certain distinct advantages and limitations of employing herbicides to control aquatic weeds in comparison to their physical control. One must weigh the two properly before opting for it in a given set of situations.

(a) Advantages

i. Herbicides control weeds quickly and, in many cases, they hold them back much longer than their physical removal.

ii. Herbicides may reach weedy situations that are inaccessible to physical or other methods of aquatic weed control.

iii. Large areas of aquatic weeds can be treated with herbicides with much greater ease than the manual or mechanical methods.

iv. With increasing cost of labour and fuel, the herbicide use is often more economical.

v. Herbicidal control minimises several labour quarrels.

(b) Limitations

i. Herbicides may impose certain water use limitations for considerable time after their application to the waterbodies.

ii. Herbicide use requires greater technical understanding on the part of the user than physical methods.

iii. Herbicides sink the weeds fast in situ. Their bulk decomposition can cause deoxygenation and eutrophication of the waterbody. Zonal treatment of the waterbody, however, can minimise such adverse effects of herbicides to a great extent.

iv. Spray and vapour drifts of herbicides may cause concern to the non-target organisms.

7.2: PRECAUTIONS IN THE USE OF HERBICIDES

1. Herbicides should not be used when non-chemical techniques offer alternative options.

2. No herbicide should be used where there is basis to believe that the water quality will be degraded and/or certain hazards exist which will unnecessarily threaten fish, wildlife, or other components of the environment, directly or indirectly.

3. Large scale, non-specific applications should not be made of any herbicide.

4. Before any herbicide is applied to a waterbody, all owners adjacent to the treatment area should be notified.

5. After the treatment, the water and food from the treated waterbody should be used strictly in accordance with the directions provided by the herbicide manufacturer or any local scientific body.

6. A contingency plan should be developed for herbicide storage areas in order to prevent their spill outside the storage places and take remedial action if the spill did occur.

7.3: COMMON AQUATIC ALGICIDES AND HERBICIDES

7.3.1: Algicides

(i) Copper compounds

Copper sulphate pentahydrate (CSP), commercially known as blue-stone, is the oldest and still the most popular copper algicide, the world-over. Its low cost is the key to its popularity. Besides, it is free from any serious application or other hazards. It contains 25% Cu^{++}. CSP will control waterbloom at 0.1 ppm, common scum algae at 1.0 ppm, and *Chara* spp. at 5 ppm or less. Of the common scum algae, *Pithophora* spp. are difficult to control with CSP but *Cladophora* and *Hydrodictyon* fade to greyish-white colour within 2-3 days after treatment. For obtaining best results the water should be soft, free of suspended soil particles, and at 16°C or higher.

Copper compounds are recommended for use in fishery lakes and ponds as well as in irrigation streams. CSP is of low toxicity to man and his domestic animals. Its safety limits are 2.3-12.0 ppm Cu in human drinking water and up to 100 ppm Cu in animal drinking water. As such, CSP is widely used for treating drinking waterbodies and storage tanks.

Copper sulphate is corrosive to common farm sprayers. Therefore, one must use fibreglass, red-brass, or copper equipment to apply it.

Copper oxychloride is another Cu-based algicide, suited for treating irrigation, recreational, and fishery waters. It is easier to apply and less corrosive to the containers than CSP.

(ii) S-Triazines

This group of herbicides includes three algicides, namely, simazine, ametryn, and terbutryn. Terbutryn has been found 64 times and ametryn 25 times more potent than simazine in this respect. Their algicidal concentrations are: 0.05 to 0.1 ppm of ametryn and terbutryn and 0.5 to 1.25 ppm of simazine. The three triazines are permitted for use in fisheries.

The s-triazine compounds are unsafe for treating irrigation and drinking waterbodies. Even the flow of water from a treated waterbody to agricultural lands is not permissible.

(iii) Dichlone

Dichlone is an effective algicide. It finds its major use in recreational waters. Dichlone will control waterbloom at 0.02-0.05 ppm and algal scums at 0.15 ppm. Another potential use of dichlone is in the control of *Chara* and *Nitella* in paddy fields.

Dichlone may sometimes irritate skin. It is toxic to fish and the treated water is unfit for drinking.

(iv) Diuron

Diuron forms an algicide, as effective as simazine. It is safe to fish. Effective rates are 0.2-0.4 ppm (or 0.5 kg ha^{-1}) against macroalgae. Waterbloom is reduced to 50% with 0.02 ppm diuron.

(v) Hexazinone

It is found effective in controlling most of the macro-algae at 0.5 ppm, while *Vauchiria* sp. is controlled at 1.0 ppm. At the present, *the use of hexazinone is restricted to non-fishery waters and artificial ponds and lakes.*

(vi) Other algicides

Nabam, Brestan 10, *Brestan* 60 and nitrofen have proved effective in controlling *Chara* and *Nitella* spp. in rice fields.

Aquashade, a blue dye, at 1-3 ppm is reported to inhibit the photosynthesis of green and blue-green algae.

H_2O_2 holds potential for controlling waterbloom caused by *Anabaena* sp. in aquaculture. It sensitises the alga to subsequent photo-injury.

Paraquat has been reported to control scum algae in tanks but it encouraged phytoplanktons (Gupta, 1976).

7.3.2: Aquatic Herbicides*

(i) Acrolein (acrylaldehyde)

Acrolein is a non-selective, contact herbicide found highly effective against submersed weeds (and algae). A wave of acrolein destroys all underwater vegetation although the perennial weeds, e.g. *Potamogeton pectinatus* re-sprout immediately after the treatment. Therefore, 2-3 staggered applications of acrolein are required to keep the water free of weeds through the season. (Besides plants, acrolein will kill snails and mosquitoes, many of which act as intermediate hosts of important human diseases).

Acrolein is widely used in several countries in the fast flowing waters *e.g.* canals, rivers, and flood control waterways as a quick clean operation. Irrigation and recreational waterbodies are particularly suited to treatment with this herbicide. Being volatile, it leaves the treated water soon. Liquid acrolein can be siphoned from barrels into the flowing waterbodies from overhead bridges and other convenient points. In general, its 1-2 barrels are applied every 6 km distance. In confined waterbodies the effective rates are 15-80 ppmv.

Despite its high effectiveness, there are certain basic limitations in the use of acrolein as an aquatic herbicide. In the first instance ***it causes massive fish kill and people have to be warned not to eat the dead fish. Also, acrolein has a pungent smell and it fouls human and cattle drinking waters.***

Sometimes the acrolein-treated irrigation water may injure crops although this aspect is not serious when the water has travelled considerable distance before reaching the fields. *Acrolein is irritating to eyes and toxic to inhale, therefore, the workmen must wear proper gas masks and gloves.*

In fisheries, acrolein may find its use in the renovation of abandoned ponds and lakes. Acrolein is also used for slime algae control in paper industry.

(ii) Amitrole and amitrole-T

Amitrole is used to control weeds on ditchbanks as well as in ditches. Amitrole-T is more mobile in plants than amitrole because of which it reaches deeper into the rhizomes and tubers of the weeds. Common application rates against *Typha* and *Phragmites* spp. are 5-15 kg ha^{-1}.

Note: *1. *Specific uses of herbicides in pisciculture and irrigation systems are detailed in Chapters-9 and 10.*

At 1-1.5 kg ha^{-1} amitrole is sometimes also considered for the control of *Eichhornia crassipes* in fish ponds because of its safety to fish.

Amitrole-treated water is largely unsuited for domestic and agricultural purposes.

Amitrole is a mild skin irritant and it is somewhat corrosive.

(iii) Ammonium sulphamate (AMS)

Ammonium sulphamate, also sometimes quoted as ammonium amino sulphate, is a highly effective, shoot mobile, ditchbank brush killer. It is suited both for the foilar and cut-stump treatments of brushes. AMS, in comparison to other brush killers, is free from problems of vapour drifts. Also, its spray drift is safe to fish and wild life though ***it can injure the nearby crop plants badly.***

AMS is corrosive to brass and copper containers.

(iv) Anhydrous ammonia

At 10 ppm or more anhydrous ammonia (NH_3) is phytotoxic to many a submersed weed and waterbloom. It is applied in confined waterbodies as liquor ammonia which maintains high pressure inside its container. The submersed weeds like *Hydrilla* and *Najas* are completely disintegrated within about five days of the treatment. The dead weeds are then collected physically. A major avenue of using anhydrous ammonia is in the renovation of uneconomic fish ponds. Its application to such ponds will destroy all vegetation as well as fish and other fauna. After some interval one could make a fresh start by introducing high quality fingerlings.

In very large fishery lakes, anhydrous ammonia can also be used safely for destroying submersed weeds by treating only 10-20% of its water area at a time which will permit the fish to move to the untreated zone (Sec. 7.4).

Ammonia departs from the treated water soon after damaging the weeds, leaving it safe for its varied uses.

(v) Aromatic solvents

Aromatic solvents are variable mixtures of substituted benzenes, chiefly xylene (= dimethyl benzene). To be effective as herbicides their aromatic content should be at least 85%.

Aromatic solvents being immiscible with water, their commercial formulations carry 2-3% of a suitable emulsifier whence these are known as *emulsified solvents*. The emulsified solvents are used in like situations and with **similar limitations as acrolein.**

(vi) Buthidiazole

Buthidiazole is found a very effective herbicide against annual and perennial ditchbank grasses and sedges. It can also be used on these weeds in drainage ditches. Its effective dose is 7.5 kg ha^{-1}. Weeds are held-back for about five months.

(vii) Chlorsulfuron

It is a potent aquatic herbicide. It is found effective against a number of difficult weeds like *Salvinia, Eichhornia, Pistia,* and *Myriophyllum* spp. at 40 g/ha or less. *Its residue problems in water should be duly considered.*

(viii) Dalaopn

Dalapon is a foliar, systemic herbicide, active on robust grasses and grass-like weeds infesting the ditchbanks and drainage ditches. It has been widely used against *Phragmites* and *Typha* spp., either straight or in combination with amitrole. With 2,4-D and silvex, dalapon forms a wide range ditchbank herbicide. Effective rates of dalapon are 10-30 kg ha^{-1}. Often its split application at about 10 days' interval is found more systemic than its one time application. *Typha* and like robust weeds are most susceptible to dalapon at their bloom stages.

Dalapon may irritate skin, eyes, and respiratory tract.

(ix) Diclobenil

Diclobenil is a powerful inhibitor of subterranean weed propagules. It is, therefore, largely used for the exposed bottom treatment of ditches. In confined waterbodies it can also be used for their through-the-water treatment after an initial mechanical weeding operation to prevent the regrowth of perennial weeds for considerable time.

For the exposed bottom application of dichlobenil the canal is drained and the herbicide granules are incorporated into the hydrosoil. The water is restored in the canal five weeks later and it is commissioned for public use after another 24 hours. Its effective rates are 5-10 kg *a.i.*/ha.

The through-the-water treatment of lakes and ponds with dichlobenil prohibits their irrigation and domestic uses for long period, although livestock could be allowed to drink water two weeks after the treatment and fishing could begin eight weeks later.

The common application rates of dichlobenil are 5-10 kg ha^{-1} for the exposed bottom treatment and 10-15 kg ha^{-1} for its through-the-water treatment. The objective should be to maintain 1 ppmw dichlobenil in water. A wide range of emersed and submersed macrophytes are effectively controlled with dichlobenil.

Dichlobenil is highly volatile in its liquid form, therefore, its granules are used in field operations. Its slow release granules are also available for the localised treatment of waterbodies.

Dichlobenil is a short persistence herbicide, of low toxicity to wildlife and fish.

(x) Diquat and paraquat

The two quaternary ammonium compounds are non-persistent dessicants of green weeds. Diquat bromide has been widely used in the USA, Europe, and elsewhere for the control of certain submersed weeds. The effective rates are 0.25 to 1.5 ppm. A mixture of diquat (or paraquat) and CSP (or Cu chelates) has proved particularly effective against *Hydrilla verticillata*. Diquat will also control certain algae but not *Chara, Cladophora*, and waterblooms. Diquat is also effective against *Pistia stratiotes, Lemna* spp., and *Eichhornia crassipes*.

In variance with diquat, paraquat is primarily used against ditchbank and shore weeds, besides against *E. crassipes*.

Diquat and paraquat find their ready use in treating irrigation and recreational waterbodies. The use of diquat, but not of paraquat, is also permitted in fisheries. These herbicides are not useful in areas where waterbodies are subjected to frequent stirring by wind to cause suspension of sediments which adsorb these cations. Also, ***the treated water is unsafe for drinking.***

(xi) Diuron

Diuron is a good soil sterilant for treating ditchbanks and drainage ditches. Also, in confined waterbodies it forms an effective preemergence herbicide against *Potamogeton, Najas* and *Ceratophyllum* spp. The effective rates are 10 kg ha^{-1} as soil sterilant and 1-2 ppm for the treatment of the water. *It may cause irritation to the eyes and skin.*

(xii) Endothall

This is primarily a shoot-active, contact herbicide. It is used in confined waterbodies for the control of a variety of submersed weeds, particularly, *Potamogeton, Ceratophyllum, Hydrilla, Cabomba*, and *Lemna* spp. besides macroalgae like *Chara* spp. While in Europe the use of endothall is discouraged because of its **high mammalian toxicity,** in the USA it is one of the most popular aquatic herbicides because of its high efficacy on weeds and short persistence (7-15 days). Although endothall and its salts are not hazardous to fish life, yet their use in fish ponds surely warrants some careful judgement. Endothall is synergistic with silvex and the combination is marketed as `Aquathol plus' for obtaining wide spectrum weed control.

Endothall irritates skin and eyes and it is toxic to mammals. To provide safety to operators, as far as possible, only its granules should be used. When it is essential to use liquid endothall, one must avoid breathing in its spray mist.

(xiii) Fenac

Fenac is a persistent, root-translocated herbicide. In waterbodies it is used for sterilising hydrosoil around docks, property water fronts, backwater areas, and lake margins. Several submersed weeds, including *Potamogeton, Myriophyllum, Elodea*, and *Najas* spp., are controlled with fenac. In specific cases it is also used to renovate stale fish ponds and to treat exposed bottoms of irrigation canals. Common rates against perennial weeds are 10-20 kg ha^{-1}. Weeds may be controlled for a year or longer.

Fenac-treated water is unsuitable for drinking and irrigation purposes (except in sugarcane fields).

(xiv) Glyphosate

Glyphosate is a non-residual, highly shoot-mobile herbicide. It is used for the control of difficult ditchbank weeds and brush. Mowed bankweeds should be allowed to regrow to 20-25 cm height before the treatment. Robust weeds like *Phragmites communis* growing in shorelines, ditches, and streams can also be carefully treated with it. Common rates are 1-3 kg ha^{-1} in up to 300 l water. Of recent, glyphosate is found particularly promising in the control of *Nymphaea* spp.

(xv) Maleic hydrazide (= MH)

MH is a plant growth retardant. It is used to defer the growth of mowed grasses on ditchbanks to provide a cover of low growing sod. Sometimes 2,4-D is mixed with MH to obtain simultaneous control of broadleaf weeds.

(xvi) Silvex (fenoprop)

Silvex is a shoot-translocated herbicide. It finds its major use in the control of difficult weeds like *Alternanthera philoxeroides, Nymphaea* sp., and *Nelumbo nucifera* in fishery and recreational waters (see chapter 9). Also, it is used for the destruction of ditchbank brushes and phreatophytes. Against water weeds, its effective concentration in water is 0.5-4.0 ppm, maintained for a period of 2-3 weeks. Silvex proves ineffective in hard waters. It is formulated as K salt and low volatile esters, both in liquid and granular forms.

The silvex-treated water is rendered unfit for its potable and agricultural uses for considerable period. It is irritating to the eyes and may be absorbed through skin.

(xvii) Sodium arsenite (Na AsO_2)

Sodium arsenite, despite its ***high mammalian toxicity***, has been widely used in U.S.A. to control submersed weeds and algae growing in confined waterbodies. There it is considered the best, all-round and cheap aquatic herbicide cum algicide when the gains from freedom from aquatic weeds outweigh its shortcomings. The effective rats are 2-10 ppmw, depending upon the limnology and temperature of the water. ***The arsenite-treated water becomes unsuitable for irrigation and bathing for considerable period after the treatment*** but it becomes potable for human beings 15 days later and for animals only 3 days after its application.

(xviii) 2,4-D

2,4-D is a translocated herbicide, effective on herbaceous broadleafs and sedges. In aquatic environment it finds its major use in the control of floating mats of *E. crassipes*. Its application rates are 1-8 kg ha^{-1}. Besides, it forms a component herbicide against ditchbank weeds. Less frequently, pellets of 2,4-D (BEE) may be used in confined waterbodies to curb submersed weeds like *Myriophyllum, Ceratophyllum*, and *Najas* spp. Diglycolic acid is found to activate 2,4-D against these weeds. The effective concentration of activated 2,4-D for treating submersed weeds is 2 ppmw.

Do not apply 2,4-D on choppy waters. Apply it when the weeds are actively growing but before they have formed dense mats.

2,4-D is also suggested for aquatic weed control in Canadian maritime oyster beds.

A major drawback of 2,4-D as an aquatic herbicide is its ***phytotoxic spray drifts*** *often reaching the nearby crops and forest plants. Floods from treated waterbodies may also* *be hazardous to agricultural and forest lands. Aerial application of 2,4-D is particularly prone to the spray drift accidents.*

2,4-D treated water is largely unfit for irrigating crops *although it is good enough for recreational purposes. In fishery waters 2,4-D may impart temporary, unpleasant flavour to fish, particularly when its ester formulations are used.*

7.4: LOCALISED CONTROL OF AQUATIC WEEDS

It is generally accepted that in many waterways total vegetation kill can severely disrupt the local ecosystem and may trigger sudden deoxygenation and heavy eutrophication. The released load of nutrients, in turn, may lead to the development of waterblooms. Several other inter-connected, adverse effects of such an action can be expected. In variance with it, when aquatic weeds are controlled in a localised area of a large waterbody at a time, disturbance in the ecosystem is minimised and the amenities of the waterways are preserved. Against floating and emersed weeds it can be achieved simply by treating the desired region of the waterbody at a time. But localised control of submersed weeds is somewhat difficult since all conventional herbicide formulations tend to disperse rapidly from the treated region of the water into the untreated areas. This also dilutes their effectiveness on the target vegetation. To overcome such a difficulty, now slow release granules, pellets, wettable powders, and viscous formulations of herbicides, which resist dispersion, have been made available.

The **slow release dry materials** are useful only for treating confined or slow moving waterbodies. Their slow release feature renders them additionally safe to fish and other aquatic fauna. For instance, the monoamine salt of endothall is normally toxic to fish at rates at which it will control *H. verticillata* and some other submersed weeds. But its slow release pellets sold as `Hydout' can bring about good, localised weed control without much harm to fish. *Dichlobenil* is another herbicide that is now made available in slow release and granular formulations *e.g.* Casoron C-SR and Casoron G, which release their toxicant slowly in hydrosoil. Weeds with well developed root system, like *Potamogeton natans* and *P. amphibium* absorb the slowly released toxicants fast and get killed. **In fishery waters one should not treat more than 10-20% of the total water area with dichlobenil granules at a time.**

Diquat and copper sulphate have been incorporated in **slow release capsules** under the trade names `Envirocap D' and `Envirocap C', respectively. These capsules provide, under certain conditions, season-long control of susceptible, submersed weeds and algae, respectively. Floating, slow release herbicide formulations, using light volcanic glass, PVC, etc., are also available in some countries for specific situations.

In certain cases wettable powders with slow release features have been formulated to obtain localised weed kill with simultaneous improvement in safety to fish and water fowl. `System E' containing endothall, is one such wettable powder. Its particles sink and settle as dust on the intercepting vegetation and the released herbicide is absorbed slowly by the vegetation for hours in calm and low turbulent waterbodies.

Copper sulphate has likewise been formulated as slow release wettable powder, chiefly to prevent its rapid precipitation in hard waters. Its one such formulation is System M. It contains 55.8% Cu and is used to control algae more effectively than CSP crystals. It can be combined with system E (containing endothall) to control algae and submersed weeds in one operation.

The slow release, viscous formulations of herbicides have been produced for use particularly in the fast moving waters. *Diquat-alginate* is one such viscous formulation of diquat. It has the appearance and viscosity similar to black treacle and contains diquat bromide and sodium alginate in different proportions. Some formulations of diquat-alginate contain also sodium hexametaphosphate, a Ca sequestering agent. It is applied to the waterbody through a 2 mm nozzle fitted to a specially modified knapsack sprayer.* It then sinks through the water in the form of strings because of which it is sometimes also known as "fibrous formulation". The diquat strings land on the submersed weeds below and stick to them.

The efficacy of diquat-alginate on weeds depends upon the Ca content of the waterbody. Both, too high and too low a Ca content will reduce its effectiveness on weeds.

7.5: DRIFT CONTROL SYSTEMS

Although granular herbicides avoid the possible hazards of spray drifts, these have their own limitations. The invert and bivert emulsion systems offer better alternatives to the granules in this respect. **Invert Emulsions** (Water-in-Oil) are now widely used to avoid spray drifts of particularly the chlorophenoxy type of herbicides. 'Instemul DA-120' is one such invert emulsion of amine 2,4-D. It is a viscous fluid and is sprayed through the bifluid system. An additional advantage of an invert emulsion system is that it tends to concentrate on the submersed vegetation in the treated area, leaving the untreated area comparatively unpolluted for fish.

Note: *1* For example, JETPACK 425 of Solo Power Equipment Ltd.*

2. Alginate (Na) = a gummy, nitrogenous organic compound obtained from seaweeds

The **Bivert Emulsion System,** is a double invert emulsion where the water phase is sandwiched between a central core and an outer layer of oil (O/W/O). Herbicide may be embodied in any one or both of these phases. The bivert emulsion is injected into the waterbody from the trailing end of a boat. It descends through the water as almost microscopic droplets with a consistency and adhesive texture of mayonnaise, and adheres to the submersed vegetation below. The herbicide is released gradually in close proximity to the weeds. The bivert system also saves herbicides like CSP, diquat, and paraquat from their inactivation in turbid waters.

7.6: METHODS OF APPLICATION OF HERBICIDES TO WATERBODIES

7.6.1: Ground Application

Patches of weeds in small waterbodies can be sprayed or treated with herbicide granules from the hand-rowed boats. Pre-pressurised, knapsack sprayers are very well suited to treat aquatic weeds since these keep both hands of the operator free. In narrow waterbodies instead of treating the weeds from within the waterbody, tractor-mounted sprayers can be operated from the shores. In wide waterbodies portable boat sprayers are used. A **boat sprayer** comprises a piston pump and a 1-3 H.P., air-cooled engine mounted on a motor boat. The pump is fitted with a dual `T' intake to simultaneously pump in water from the waterbody and the toxicant from the tank (Plate.7.1). A garden hose extends from the outlet to the spray gun. The spray range of a boat sprayer is usually 8 to 20 m with a discharge of 18 to 160 1pm and spray pressure of 4 to 5 kg cm^{-2}. For spraying the floating weeds, the spray gun is usually fitted with flat spray type of nozzles. When submersed weeds are to be treated from a boat sprayer, the spray gun is replaced by a suitable hard pipe having several, small holes, spaced 30-50 cm apart. Alternatively, several long hoses can be used in place of a multi-holed pipe to inject the herbicide well below the surface of water. This technique of herbicide application is called **bottom placement**.

There is another device for incorporating herbicides evenly in waterbodies for the control of submersed weeds. This is called **boat bailer** (plate. 7.2). It comprises a tank to carry the herbicide solution (or emulsion) of approximately 5% concentration. The outlet of the tank is connected to the boat bailer through a flexible tube and regulator. The boat bailer, in turn, is attached to the out-board motor near the propeller so that as the boat moves ahead, the solution is released into the waterbody by gravity and is mixed with water by the propeller-induced turbulence. The discharge of the solution should be increased when passing through deeper parts of the waterbody to maintain uniformity of the herbicide concentration.*

---

3. O = Oil; W= Water

Plate 7.1: A boat sprayer showing its water pump meant for mixing aquatic herbicides in the tank with water drawn from the lake before its application to the target aquatic weeds with a spray gun. (Courtesy: Lopinot, 1968)

Plate 7.2: A boat bailer devised to release a concentrated herbicide solution/emulsion directly from the tank into the waterbody through a flexible tube attached to the boat bailer to treat the submersed weeds. (Courtesy: Lopinot, 1968).

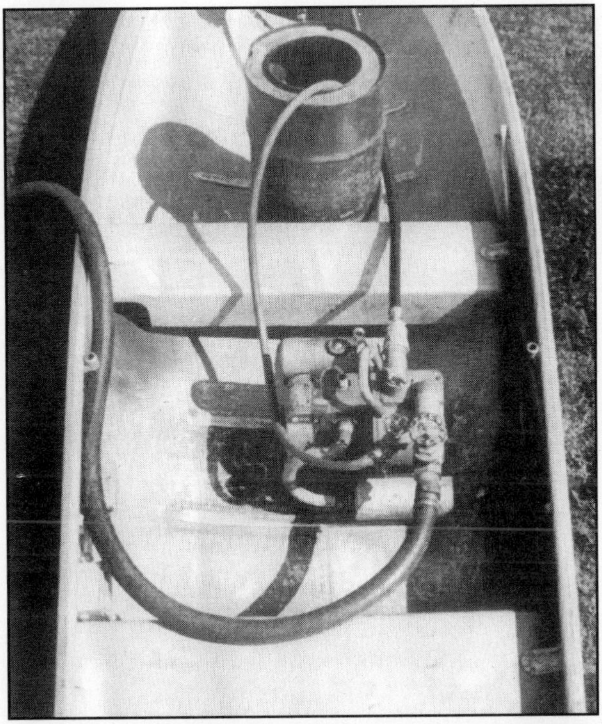

Plate 7.1

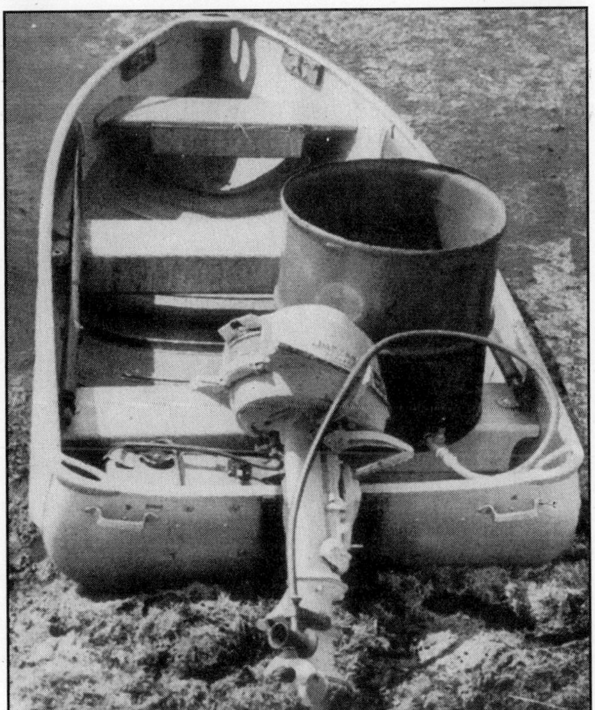

Plate 7.2

Plate 7.3: A boat sprayer in operation somewhere in USA against waterhyacinth.

Plate 7.4: A newly laid drainage channel at Kota treated with monuron. Although attractive, the treatment has its own limitations. (Photo-author).

Plate 7.3

Plate 7.4

Normally, during the operation the speed of spray boats and boat bailers is 3.5 to 5.0 km hr-1. But it is not uncommon to come across weed growths so thick that the spray boats have to be rowed or dragged with ropes. In such events the pumps are required to be recalibrated to account for the loss in work speed. To obviate this, three new application devices have been developed for treating the overweedy waterbodies. These devices are: (i) Hoovercraft (ii) Spray Raft, and (iii) Weed Wiper.

A **hoovercraft** is a strong rubber boat powered by one or two aeroplane propellers placed on the rear end, instead of the conventional underwater propellers which are easily caught in weeds. The hoovercraft continues to move even in badly weedy waters, without any difficulty.

A **spray raft** is a very simplified version of hoovercraft. It was developed in Australia to facilitate spraying of floating mats of weeds like *Eichhornia, Pistia*, and *Salvinia* spp. It consists of an inflated inner tube of a truck tyre and an ordinary irrigation sprinkler. The raft can easily be made to slide over the weed mats with the help of a light rope tied to it and stretched across the area to be sprayed. One man can easily pull the raft, treating about 24 m wide swath at a time. The herbicide mixture is pumped through a hose to the sprinkler. This hose also serves as a tow rope for the retrieval of the raft after each crossing. For easing the operation the spray hose and the towing rope should be attached from underneath the raft.

A **weed wiper** is a tractor-mounted, hydraulic arm for smearing shoot-active herbicides, *e.g.* glyphosate and paraquat, on the foliage of tall, emersed weeds in ditches, streams, and in some other difficult to approach situations.

The herbicide granules are applied by broadcast on the surface of water, as also on ditchbanks, with **rotary seeders** and **granule blowers.**

7.6.2: Aerial Application

Vast infestations of aquatic weeds are more conveniently sprayed with herbicides from air than from the ground. This is particularly true where the very approach to the weeds from the ground is difficult. Helicopters are preferred over other aircrafts for spraying waterways because of their great maneouverability. Thus far, aerial application of herbicides has been mostly done against *E. crassipes* and *Typha* spp. in certain advanced countries. A helicopter can cover up to 200 ha water surface a day. A major bottleneck of aerial application of herbicides is its ***drift hazards*** to nearby forest and field crops, particularly from the phenoxy type of herbicides. However, this can be minimised by several mechanisms already discussed earlier.

7.6.3: Special Herbicide Application Techniques

(a). ***Copper sulphate:*** For its application in flowing waterbodies, CSP crystals are contained in nylon mesh or burlap bags and suspended, 1/3-1/2 submersed in water. Alternatively, the crystals can be simply shovelled into large canals from above a concrete apron. In confined waterbodies, on the other hand, a burlap bag containing CSP crystals is dragged behind a motor boat covering the intended water surface, systematically.

(b). ***Emulsified solvents:*** The emulsified solvents are pumped or injected beneath the water surface under pressure of 3.5-28 kg cm^{-2}. It is done in two ways. First, in turbulent waters the emulsified solvent is directly pumped from the holding tanks at a point of moderate turbulence. For this, a low volume pump unit equipped with a small orifice nozzle boom is used. Second, in non-turbulent waters, high volume, centrifugal pump unit with a fire-hose nozzle at the delivery end is used. Here the pump has a double suction line. The main 10 cm wide suction line draws water from the channel into the belljar of the pump, while the second, about 3 cm wide line draws the emulsified solvent into it. In the belljar the two are mixed before the diluted solvent is discharged in water through the firehose.

(c). ***Anhydrous ammonia*:** Liquid ammonia is applied under its own pressure direct from the holding cylinders through a long rubber tube ending in a short metal tube. A meter is provided to read the quantity of ammonia released. The liquid is delivered about 60 cm beneath the water surface. Being lighter than water, ammonia tends to move up and concentrate in the top 60 cm depth of water where the weeds are most dense (Ramachandran, 1960).

7.6.4: Treatment Of Hydrosoil

When it is possible to drain a waterbody completely and expose its hydrosoil, soil-active herbicides like **fenac** and **dichlobenil** can be applied to it. The treatment will destroy many weedy rhizomes, roots, and tubers and provide extended period control of the submersed and emersed weeds. The exposed bottom treatment is made on the ground area basis and it is incorporated physically in to the hydrosoil. The treated bottom is left undisturbed, well exposed to sun for a prescribed length of time. The water is then let in to fill the waterbody to a desired level and it is commissioned 1-2 days later.

7.7: CALCULATING HERBICIDE REQUIREMENTS

Waterbodies are so variable in shapes and sizes that no single formula can be used to determine the quantity of herbicide needed to treat them. Still, persons engaged in aquatic weed control work should make themselves familiar with some basic calculations involved in determining the herbicide requirements for the confined and moving waterbodies.

The surface floating and emersed weeds, irrespective of the shape of a waterbody, are treated on the basis of the area they infest. The herbicide doses for such weeds are, therefore, given in terms of kg ha^{-1} which are easy to calculate. But when an aquatic herbicide is to be applied for the control of submersed weeds in a *confined waterbody*, one has to calculate its quantity to produce a recommended concentration in specified column of water. For this, one needs to know the volume (V) or weight (W) of water to be treated. The volume (V) of water in cu m is determined using the formula:

$$V = A \times D,$$

where 'A' is the area of water surface in sq m and 'D' the average depth of water column in m.

While the area of the waterbody can be determined using an appropriate geometrical formula depending upon its shape (circular, rectangular, triangular, etc.), the depth of the water column can be measured with a calibrated, long pole or weighted rope. Several readings must be taken in straight lines, 9 to 15 m apart. As a working rule, the average depth of a small lake is equal to half the maximum depth measured.

Once the volume (V) of water under treatment is computed, the quantity of herbicide (Q) required to produce a desired concentration (C) is given by:

$$Q \text{ (in kg)} = V \text{ (in cu m)} \times 0.001 \times C \text{ (ppmw)}$$

Sometimes, the volume of water is determined in ha m (instead of cu m). In that case,

$$Q \text{ (in kg)} = V \text{ (in ha m)} \times 10.11 \times C \text{ (ppmw)}^{*}*$$

In ***flowing waterbodies*** the quantity of herbicide applied for the control of submersed weeds is calculated on the basis of their discharge in cumsec (cubic meter per second). This can be determined from the measurement of average width (W), depth (D), and flow velocity (U) of the ditch. Then discharge of water in cumsec is calcuted as:-

$$W \times D \times U \times 0.9,$$

---

* ***Note:*** *1. An interesting rule of thumb for treating confined ditches is that an application of 500 g of herbicide to each sq m cross section of a km long ditch produces a concentration of 0.5 ppmw in the water they carry at full capacity.*

where 'W' and 'D' are in m and U in m sec^{-1}.

Table 7.1: Ready Reckoner For Herbicide Required To Treat Waterbodies Of Variable Surface Areas And Depths To Obtain One Ppmw Concentration Of The Active Herbicide.

Average depth of water (m)	Surface area of water 1 ha	2 ha	3 ha	4 ha	5 ha	0.1 ha
1.0	10	20	40	40	50	1.0
1.5	15	30	45	60	75	1.5
2.0	20	40	60	80	100	2.0
2.5	25	50	75	100	125	2.5
3.0	30	60	90	120	150	3.0
3.5	35	70	105	140	175	3.0
4.0	40	80	120	160	200	4.0

The dose of the herbicide for treating flowing waterbodies is given as kg/cumesc discharge. In the case of emulsion solvents and acrolein, the delivery rates (or the application time) are also indicated.

7.8: FIELD EXPERIMENTATION WITH AQUATIC HERBICIDES*

Submersed aquatic weeds pose certain peculiar problems for the experimenter which are not encountered in terrestrial weed control research. In the first instance there are practical difficulties in observing the plants and making their quantitative assessments under water. Then there are problems relating to direct and indirect treatment effects on the aquatic environment. This makes a comparison between treatments much more complex and, sometimes, impossible. This section reviews the techniques currently in vogue in conducting field research on herbicidal control of aquatic weeds and ways to minimise the above cited difficulties.

7.8.1: Choice Of The Experiment Site

The choice of the site depends upon not only the type of experiment required but also upon the local conditions of water use and water flow. Priority must be given to avoid any risk to the irrigated crops and local fishing interests, and to secure the cooperation of all concerned parties. Further, the site under consideration must be as far as possible uniform in weed population and water depth. It must be accessible to the experimenter

---

* ***Note:*** * *Adapted from T.O. Robson and P.R.F. Barrett (1972), BWCC, England.*

and, preferably, not to the public. Assurances should be obtained that the experimental area will not be interfered with for the period of the experiment to allow for adequate assessments. This may include follow up weed growth measurements in the subsequent seasons.

7.8.2: Choice Of Experimental Design

The size and shape of the waterbody will determine to a large extent the type of experiment. A large, uniform area such as a lake may lend itself to sub-divisions for a replicated experiment, whereas drainage ditches, however stagnant they may appear to be, are always liable to flood and are generally suitable only for simpler designs.

(a) ***Replicated experiments*:** In aquatic weed research it is not possible to use the more advanced experimental designs generally adapted in agronomic experiments. In certain circumstances, at most, randomised block design (RBD) may be applicable: if this design is adopted, it is necessary to somehow isolate plots and avoid cross-contamination of treatments within a replication. In lakes, one way to do so is to pre-fabricate enclosures, constructed with wooden frames supporting the polythene sheeting. The corner posts of the frames must be pointed and protrude about 15 cm below the lower cross bar so that they can be driven into the bottom until the cross bar and polythene `skirt' is buried in the mud. This is essential to form an adequate seal and to prevent the loss of the herbicide. Satisfactory enclosures have been made out of 5 X 5 cm timber but in U.S.A. less robust frames have been used. The enclosures are usually 4 m^2 each. These can be positioned with spaces left between them to prevent cross-contamination. Such enclosures provide acceptable experimental conditions for few weeks but no longer. It is so because with prolonged isolation of water inside enclosures from the main waterbody, seasonal changes in biological populations and solutes will differ and eventually conditions within the enclosures will become atypical and unreliable for comparison with the other enclosures. However, for short term experiments (up to 2 months in summer) this type of layout is valuable.

Attempts have been made to replicate treatments in separate drainage ditches and this can be successful if the weed populations were similar and careful records were made before application of the treatment. The main risk then is an unequal flow of water which will affect herbicide concentration and, therefore, the duration for which the plants are exposed to the chemical. Regular monitoring of residues will help, but it will not be possible to apply any statistical analysis of variance to the results. Replication in the same channel may be done by inserting barriers made of polythene sheeting or hardboard across the water course and pressing their bases well into the bottom mud. In most channels, however, it is impossible to stop water movement entirely in this way and a large gap (50 m or more) should be left between plots. Residue data should be obtained from the treated water to ascertain the herbicide concentration at intervals throughout the experiment.

(b) *Unreplicated experiments*: This type of experiment may involve several treatments and, therefore, barriers or enclosures are normally still necessary. However, by reducing the number of plots it is often possible to increase their size, thus avoiding the problem of atypical conditions associated with small plots. Where experiments are unreplicated there is no need to randomise the layout and the untreated controls should be placed upstream as an additional safety measure to prevent contamination. Unreplicated experiments involve the risk that results may be lost when, for instance, a plot barrier breaks. It is usually advisable to repeat the trials on a number of different sites in the same year as a precaution against this risk and to provide confirmation of the results. In fact, different locations can act as replications to the trial.

The use of large plots has important advantages over the smaller enclosures as it can generally be assumed that any difference between the biological systems in the treated and untreated water will be the result of the treatment and not induced by the enclosures. However, to be sure of this it is necessary to have an untreated and uncontaminated body of water containing the test species in close proximity to the treated plots. This will provide a check on the natural, seasonal development of the plants.

Frequently, it is not possible to insert barriers across ditches and the treatments must then be separated by the untreated buffers zones. The length of these buffers will depend upon the expected movement of water and the formulation of the herbicide being used but they should usually be no less than 100 m long. All aquatic herbicides will diffuse and move with water. The value of the experiment can be enhanced by regular monitoring of the concentration in the treated portions.

It is sometimes preferable to treat the full lake at one time without attempting to divide it with the help of enclosures or barriers. The value of the results of this kind of experiment depends upon the amount of care and effort put into the sampling and assessment by the experimenter. If done well, it can provide a large amount of valid information.

7.8.3: Assessment Of Effects On Weeds

The two most important elements for successful assessment of treatment effects on weeds are: (i) to know what plants were there at the time of treatment and which of these species survived the treatment, and (ii) to obtain evidence that the species that were killed would have survived had they not been treated. To meet these objectives it is needed to make careful records of the species distribution before and at intervals after treatment. These may be either simple observations along a ditch or more complex mapping but greater the care taken in the collection of the data more valuable are the results.

Aquatic weed populations are usually a mixture of numerous species and early in the year they are difficult to distinguish from above the water. It is, therefore, necessary to take weed samples for identification. The assessment of infestation of emersed and surface floating weeds is easy by conventional quadarate method, or even by eye estimation. But accurate estimation of submersed weeds is difficult. One way to do so is to throw from a boat either rake or grapnel,* secured to a strong rope in water. The rope should be loosened until the hooks strike the bottom. When pulled back, a representative sample of submersed weeds is usually available. Several samplings at different locations in the waterbody will give a fairly good idea of the weed species composition. But in case of mixed population of several weed species, some scientists prefer to use permanent transects and quadarates. The area that each species occupies can then be estimated as a percentage of the total transect area and plotted or sketched on a transect plan so that the position of each may be located at subsequent assessments.

At the Central Inland Fisheries Research Sub-Station, Cuttack (India) a light, cubic, metal frame of known volume is used, instead of hooks, for sampling of submersed weeds. This equipment has the advantage of obtaining numerical data on the weight of weeds per cu m of water, but it is laborious, besides being inconvenient in deep waters.

Identification of weed samples is difficult in the early growing period when only the vegetative parts are available. If there is any doubt about its identity it is advisable to mark a plant of the same species growing in the control plot so that it may be examined at the time of its flowering and fruiting stages. If required, it may be sent to an herbarium for its exact identification.

Some scientists have used regular, visual estimates of biomass on a 5-point scale at intersections of a grid to assess the effect of herbicides on submersed plants. They then compared the records by means of different sized dots on a series of maps for species and sampling dates. This gave a very good record of the changes that took place. However, it is probably more detailed than is required for many herbicide investigations.

It is often important to record the relative sizes of the plants, besides their presence. A method of doing so is by means of symbols representing the visual estimates.

The use of photometers to estimate light extinction among weeds as indirect measure of their densities has received considerable attention. It is discussed with the general problem of macrophyte sampling in the IBP Handbook No. 12 [Vollenweider, R.A. (Ed) 1969]. It is, however, not applicable to mixed species stands and is very difficult to

---

* *Grapnel = a small anchor with several claws or arms.*

operate if filamentous algae were present which tends to wrap around the photocell and disrupt light measurements.

7.8.4: Assessment Of Effects On Aquatic Organisms

Phytoplankton, zooplankton, and other invertebrate animals are often migratory and short-lived, and the populations fluctuate throughout the year. It would be necessary to build up a detailed picture of these population fluctuations over a number of years in order to demonstrate minor changes resulting from a herbicide treatment. However, only gross changes are usually of interest and these can be detected by monitoring before and after the treatment. Sampling devices have been developed by freshwater research organisations and university departments and are discussed in the IBP Handbooks referred to above.

7.8.5: Assessment Of Environmental Factors

The most important environmental factor involved in the assessment of chemical weed control in waterbodies is the level of dissolved oxygen. Deoxygenation is the main risk to fish and other animals following the use of herbicides in confined waterbodies. In a lake, a reduction in its O_2 content from 10 mg l^{-1} water to 0.44 mg l^{-1} in 8 days, following application of paraquat, was recorded.

Measurement of dissolved oxygen in the waterbodies is now a simple matter with portable dissolved oxygen meters. Their operation involves introducing an electrode into the water and reading the percentage saturation. The water temperature is also recorded simultaneously. These data can then be converted to mg of dissolved oxygen present per l of water. However, to obtain a reliable estimate, a number of readings from different points in the waterbody are required. Care should be taken to avoid placing the electrode of the oxygen meter near the hydrosoil where anaerobic conditions are normally found. Also, it is advisable to measure the dissolved oxygen of the experiment area at the same time of day on each occasion to minimise the effect of difference in the day-to-day rate of photosynthesis and consequently in the oxygen evolution. The most reliable sampling time is at dawn before photosynthesis starts, but this is seldom possible and a compromise has to be found.*

---

* *Note: 1. The choice of herbicide will depend upon the intended use of the waterbody as per Chapters 9, 10, and 11 and Table 7.3.*

Table 7.2: Susceptibility of Certain Aquatic Weeds To Herbicides At A Glance

Weed	Herbicides*
1. Algae (lower)	Copper compounds, terbutryn, ametryn, simazine, dichlone, diuron, hexazinone, and paraquat.
2. *Alternanthera philoxeroides*	Silvex, paraquat, and glyphosate (only in dried ditches)
3. *Cabomba* spp.	Diquat, 2,4-D, silvex, and endothall
4. *Carex* spp.	Dalapon
5. *Ceratophyllum* spp.	CSP, endothall, 2,4-D, silvex, and diquat
6. *Chara* (and *Nitella* spp.)	Copper compounds, 2,4-D, nabam, dichlone, brestan, nitrofen, simazine, diuron, endothall, and dichlobenil
7. *Cladium* sp.	Dalapon
8. *Cyperus* spp.	TCA, dalapon, and buthidiazole
9. *Eichhornia crassipes*	2,4-D, diquat, paraquat, glyphosate, amitrole-T, ametryn, hexazinone
10. *Egeria densa*	Diquat with endothall
11. *Elodea* spp.	Diquat, endothall, and silvex
12. *Ipomoea* spp.	2,4-D
13. *Hydrilla verticillata*	Anhydrous ammonia, diquat plus CSP, acrolein, xylene, endothall, fenac, dichlobenil, bensulfuron, and fluridone
14. *Juncus* spp.	Silvex
15. *Lemna* spp.	2,4-D, diesel oil, diquat, endothall
16. *Myriophyllum spicatum*	2,4-D, glyphosate, silvex, diquat, and endothall
17. *Najas* spp.	Endothall, diquat, silvex, and 2,4-D.
18. *Nelumbo* spp.	2,4-D, TCA, CSP, dichlobenil, and silvex.
19. *Nitella* spp.	(see Chara spp.)
20. *Nymphaea* spp.	Glyphosate, 2,4-D, TCA, CSP, endothall, dichlobenil, and silvex.
21. *Nymphoides* spp.	TCA
22. *Phragmites* spp.	Buthidiazole, dalapon, amitrole, 2,4-D, TCA, herbicidal oils, and glyphosate.
23. *Pistia stratiotes*	Mineral oil, 2,4-D, paraquat, anhydrous ammonia, diquat, amitrole-T.
24. *Potamogeton* spp.	2,4-D, chlorthiamid, acrolein, xylene, endothall, dichlobenil, and diuron.
25. *Ranunculus* spp.	Diquat
26. *Sagittaria* sp.	2,4-D, silvex, and endothall.
27. *Salvinia auriculata*	Kerosene with diuron, paraquat, PCP, hexazinone, and ametryn.
28. *Scirpus* sp.	2,4-D, amitrole, and dalapon.
29. *Spirodela* spp.	2,4-D and diquat.
30. *Typha* spp.	2,4-D, dalapon, amitrole, glyphosate, diquat, and paraquat.
31. *Utricularia* spp.	2,4-D, silvex, endothall, sodium arsenite, and diquat.
32. *Vallisneria spiralis*	2,4-D and CSP.
33. *Wolffia* spp.	Diquat
34. *Zannichellia palustris*	Endothall

Table 7.3: Average Suitability Of Certain Aquatic Herbicides For Different Uses Of Waterbodies At Their Recommended Application Rates. (S = safe; NS = not safe)

	Aquatic herbicides	Use of the treated waterbody				Mammalian toxicity during application
		Irrigation	Fishery	Domestic	Recreation	
1.	Acrolein	S	NS	NS	. S	Irritant, toxic to inhale.
2.	Amitrole	NS	S	NS	NS	Irritant
3.	Anhydrous ammonia	S	NS	S	S	–
4.	Aromatic solvents	S	NS	S	S	Irritant
5.	Copper compounds	S	S (NS for trout)	S	S	Toxic (low)
6.	2,4-D	NS	S (except slow release forms)	S	S	–
7.	Dichlobenil					
	i. Exposed Bottom troatmont	S	S	S	S	–
	ii. Through-the-water treatment	NS	S (After 8 weeks)	NS	NS	-
8.	Diquat (free of surfactant)	S	S	NS	NS	Toxic, Irritant
9.	Diuron	NS	S	NS	NS	Irritant
10.	Emulsified solvents	S	NS	NS	S	Irritant
11.	Endothall	S	S (Na and K salts)	NS	NS	Toxic, irritant
12.	Glyphosate	S	NS	NS	S	–
13.	Silvex (fenoprop)	NS	S	NS	S	Irritant
14.	Sodium arsenite	NS	S	NS	NS	Toxic
15.	Simazine	NS	S	NS	S	–

Comment: *Except for the Cu compounds, no aquatic herbicide is safe for all major uses of the treated waterbody. This limits their deployment in multifarious utility waterbodies in India and like countries. And the Cu compounds are good enough only for controlling the weedy algae in waterbodies.*

SELECTED BIBLIOGRAPHY

1. Ahmed, S.A. *et al.* 1982. Water quality as affected by waterhyacinth decomposition after cutting or 2,4-D application. Weed Res. (Japan) 27 (1): 40.

2. Alabester, J.S. and F.S.H. Abram. 1965. Estimating the toxicity of pesticides to fish. PANS (C): 11 (2): 91-97.

3. Aly, O.M. and S.D. Faust. 1964. Study of fate of 2,4-D and ester derivative in natural surface waters. J. Agr. Food Chem. 12: 541-545.

4. Anderson, L.W.J. 1976. Status of classification of aquatic herbicides. J. Aquat. Plant Manage. 14 (June): 1-3.

5. Anon. 1968. (Gramoxone-S and aerial gramoxone for the control of aquatic weeds. Focus (Imperial Chemical Industries) 4073, 6/68.

6. Bachmann, R.W. 1980. The role of agricultural sediments and chemicals in eutrophication. J. Water Pollution Contr. Fed. 25 (10): 2425-2431.

7. Baker, G.E. *et al.* 1975. Improved application techniques for aquatic herbicides. Hyacinth Contr. J. 13: 21-23.

8. Barrett, P.R.F. 1974. A spraying rig for the experimental application of herbicides to the floating leaves of water plants. Weed Res. 14: 313-315.

9. Blackburn, R.D. and L.W. Weldon. 1965. The sensitivity of duckweeds (Lemnaceae) and azolla to diquat and paraquat. Weeds 13, 147-149.

10. Bruns, V.G. *et al.* 1972. Response of several crops to six herbicides in irrigation water. Tech. Bull. 1461, Agr. Res. Sta., USDA, 29 p.

11. Bruns, V.G. *et al.* 1974. Responses and residues in six crops irrigated with water containing 2,4-D. CARC, Wahsington State University, Bull. No. 798, 8 p.

12. Butter, J.N. 1972. Water quality and chemical models. Environ. Affairs 2 (2): 382-398.

13. Cope, O.B. 1965. Agricultural chemicals and freshwater ecological systems. *IN* Research In Pesticides, Acad, Press Inc. N.Y. 115-127.

14. Cummings, J.G. 1976. Residue tolerances for aquatic herbicides. J. Aquat. Plant Management 14 (June): 4-6.

15. Demint R.J. 1970. Amitrol residues and rate of dissipation in irrigation water. Weed Sci. 18 (4): 859-864.

16. Fitzgerald, G.P. and S.L. Faust. 1963. Factors affecting the algaecidal and algaestatic properties of copper. Applied Microbiol. 11: 345-351.

17. Foster, H.S. and A.V. Coombs. 1962. The control of algae with copper sulphate. World Crops 14: 32, 36.

18. Frank, P.A. 1966. Persistence and distribution of monuron and neburon in an aquatic environment. Weeds 14: 219-222.

19. Frank, P.A. and R.D. Comes. 1967. Herbicidal residues in pond water and hydrosoil. Weeds 15: 210-213.

20. Frank, P.A. *et al.* 1961. Techniques for evaluating aquatic herbicides. Weeds 9 (4): 515-521/

21. Gangstad, E.O. 1976. Dissipation of residues of 2,4-D in water, hydrosoil, and fish. Hyacinth Control J. 14: 43-44.

22. Gaylor, J. and A. Houser. 1962. The three years' results with silvex for aquatic plant control in Oklahama. Down to Earth 18: 2-4.

23. Gupta, O.P. 1973. Aquatic weed control. World Crops 25 (4): 132-190.

24. Gupta, O.P. 1976. Chemical control of hydrilla (*Hydrilla verticillata* Casp.) (Hydrocharitaceae). Raj. Pesticides 3 (1 & 2): 18-24.

25. Hardy, J.L. 1966. Effect of Tordon herbicide on aquatic chain organisms. Down to Earth 22, 11-13.

26. Hestand, R.S. and C.C. Carter. 1977. Succession of various aquatic plants after treatment with four herbicides. J. Aquat. plant Manage. 15 (June): 60-64.

27. Hiltibran, R.C. 1962. Duration of toxicity of endothall in water. Weeds 10 (1): 17-19.

28. Hiltibran, R.C. 1963. Effect of endothall on aquatic plants. Weeds 11 (14): 256-257.

29. Joshi, N.C. and P. Paharia. 1969. Suggested herbicides against aquatic weeds. Pesticides 3 (7): 27-31.

30. Kasai, F. 1993. Effects of simetryne on the growth of various freshwater algal texa. Environ. Pollution 79 (1): 77-83.

31. Manning, J.H. and R.E. Jonson. 1975. Water level fluctuation and herbicide application. Hyacinth Contr. J. 13: 11-16.

32. Marston, R.B. *et al.* 1968. Amitrole concentrations in waters downstream from an aerially sprayed watershed sub-basin. Pest Monitroing J. 2: 123-128.

33. Mullison, W.R. 1970. Effects of herbicides on water and its inhabitants. Weed Sci. 18 (6): 738-750.

34. O 'Neal, S.W. and C.A. Lembi. 1983. Effect of simazine on photosynthesis and growth of filamentous algae. Weed. Sci. 31 (6): 899-903.

35. Parkar, C. 1966. Influence of water hardiness on the phytotoxicity of paraquat. Nature (Lond.) 212, 1465-1466.

36. Ramachandran, V. 1960. Observations on the use of ammonia for the eradication of aquatic vegetation. J. Sci. Indust. Res 19 (II): 284-285.

37. Ransom, J.K. *et al.* 1983. Behaviour of 2,4-D in common waterplantain (*Alisma trivale*). Weed Sci. 31 (6): 766-770.

38. Robson, T.O. 1967. A list of herbicides used for aquatic weed control. PANS (C) 13: 204-206.

39. Robson, T.O. 1983. Localised control of submersed aquatic weeds with herbicides. (*IN*) Recent Advances in Weed Research (*Ed.* Fletcher, W.W.) C.A.B., U.K. 171-180.

40. Robson, T.O. *et al.* 1976. Effect of some herbicides on freshwater algae. Pesticide Sci. 7: 391-402.

41. Sahai, R. *et al.* 1980. Effect of certain herbicides on the growth performance of *Spirodella polyrhiza.* Ind. J. Weed. Sci. 12 (1): 87-92.

42. Sanders, H.O. 1970. Toxicity of some herbicides to six species of freshwater crustaceans. J. Water Pollution Control Federation, 1544-1550.

43. Schultz, D.P. and E.O. Gangstad. 1976. Dissipatation of residues of 2,4-D in water, hydrosoil, and fish. J. Aquat. Plant Manage. 14 (6): 43-44.

44. Sen. A.C. 1957. The effect of 2,4-D weed killer on aquatic fauna. Sci. Cult. 23: 315.

45. Sisneros, D. *et al.* 1998. Low dose metering of endothall for aquatic plant control in flowing water. J. Aquat. Plant Manage. 36: 69-72.

46. Snow, J.R. 1963. Simazine as an algaecide for the bass ponds. Prog. Fish Culturist 25 (1): 34-36.

47. Tanaka, H. and Y. Yoshizawa. 1982. Chemical control of algae on clay flower pots. Bu.. Faculty of Agri., Tamagawa Univ. No. 22: 45-55.

48. Timmons, F.L. *et al.* 1979. Herbicide residues in agricultural water for the control of aquatic and bankweeds. (*IN*) T.L. Willbrick and G.E. Smith (*Ed.*). Agric. Pesticides and Water Quality, Iowa State Univ. Press, USA.

49. Yeo, R.R. 1967. Dissipation of diquat and paraquat and effects on aquatic weeds and fish. Weeds 15: 42-46.

50. Yeo, R.R. 1970. Dissipation of endothall and effect on aquatic weeds and fish. Weed Sci. 18 (2): 282-284.

51. Walker, C.R. 1964. Toxicological effects of herbicides on the fish environment. Water and Sewage Works, Parts I and 2, III (3), 8 pp.

52. Walker, C.R. 1971. Problems in clearance and registration of herbicides for aquatic areas. Hyacinth Control J. 10: 5-17.

53. Way, J.M. *et al.* 1971. Some ecological effects of the use of paraquat for the control of weeds in small lakes. J. Applied Ecol. 8 (2): 509-532.

54. Weldon, L.W. *et al.* 1966. Use of invert-emulsion formulations in aquatic weed control. Hyacinth Control J. 5: 12-14.

55. Wells, R.D.S. 1994. The efficacy of copper triethanolamine complex on waternet. Weed Astr. 44 (10): 4308.

56. Wingfield, G.I. and J.M. Johnson. 1981. Deoxygenation of water following use of the herbicide terbutryn, in a simulated batch culture system. Bull. Environ. Contamination Toxicol. 26: 65-72.

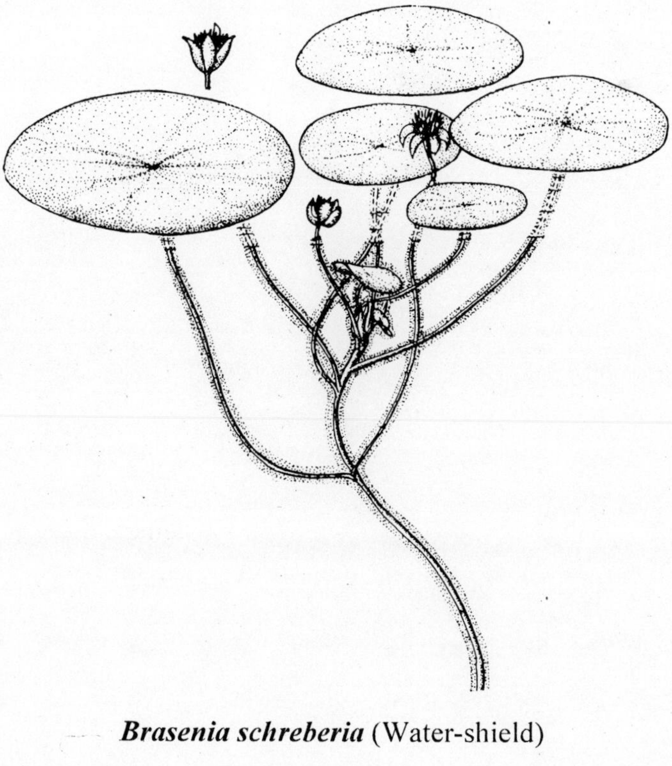

Brasenia schreberia (Water-shield)

Chapter 8

BIOLOGICAL CONTROL OF EXCESSIVE AQUATIC PLANTS

A number of noxious aquatic plants present with us today are alien species, introduced purposely by man as ornamental pond plants to decorate his surroundings, for instance waterhyacinth, waterlettuce, salvinia, and waterspinach. Also, man has introduced some plants for his aquarium needs *e.g.* hydrilla and najas. Other modes of introduction of aquatic plants into India cannot be ruled out, particularly the long distance navigation which helped an aquatic plant like alligatorweed to disperse widely. While such aquatic plants moved from their places of origin (largely South America and around) to the new areas, their native insect pests and disease pathogens were left behind which had kept these plants in check back home. In other words, the nature's biological way of limiting the populations of these aquatic plants was disturbed and the introduced aquatic plants had a free go in the new environment to infest various waterbodies in ever increasing proportions. The scientists realised the phenomenon and made first attempt to reverse it about 100 years back by introducing insect pests of specific weeds from their centers of origin. Initial success in these efforts was achieved on terrestrial weeds, starting with lantana. This approach to weed control was named as Biological Control of Weeds. Subsequently several different kinds of bioagents, apart from insects, were tried on different extraneous weeds, including the aquatic ones.

Today, the biological control of weeds is defined as the use of living organisms, including insects, plant pathogens, herbivorous fish and other animals, and competitive plants to limit their infestations. A major condition of a successful bioagent of any weed before it is released in an area is that it should not damage our crops or other economic plants even if it had to starve to death in the absence of the target weedy plant. However, in recent years there is a trend amongst the scientists to give some lalitude to this condition. Further, bioagent to be successful should be able to easily establish itself in the new environment; inflict acceptable damage to the target weed; and be free from predators, including fish.

The rising concern towards residual toxicity of herbicides used for aquatic weed control has given impetus to the scientists the world-over to intensify research on the biological control of aquatic weeds. This method is innocuous to the non-target aquatic organisms, man, and other users of the waterbodies. Also, once released, at least the arthropod bioagents are largely self propagating without the need of repeated treatments, unlike the herbicidal and physical control options, and thus prove highly economical. The biological control of aquatic weeds is a slow acting one which lets the weedy mats to sink slowly and avoid rapid putrefaction and sudden rise in B.O.D. of the waterbody (Chapter-1). Also, since in the biological control of aquatic weeds with insects the populations of the bioagent and the target weed remain in cyclic balance, the chances of rapid weed species ecoshifts are few.

Quite some specific and non-specific bioagents have been successfully discovered and used against certain weeds in different countries. Some successful examples of these are given here as under:

8.1: ARTHROPODS

Probably, the first successful attempt in biological control of aquatic weeds using insects was against alligatorweed (*Alternanthera philoxeroides*) in the USA. Between 1963 and 1970, three insect bioagents *viz*., fleabeetle (*Agasicles hygrophila* Selman & Vogt) (Fig. 8.1), alligator thrip (*Amynathrips ondersoni* Neill), and alligator stem borer (*Vogtia malloi* Pastrana) were established against this noxious aquatic weed.

These insects were imported into Florida from their native home, South America. The Florida bayous were badly choked with this weed before *A. hygrophila* was released. Today, the weed does not create any obstruction in the rivers there. Gangstad *et al.* (1975) further reported that an initial stress to the alligatorweed with low doses of 2,4-D or diquat paved way to its fast destruction by the flea beetle.

Researches on the biological control of waterhyacinth (*Eichhornia crassipes*) have been very successful with three insect bioagents *viz*., (i) *Neochetina eichhorniae* Warner, (Fig. 8.2), (ii) *N. bruchi* Hustache (hyacinth weevils), and (iii) *Sameodes albiguttalis* Warren (hyacinth moth) (Fig. 8.3).

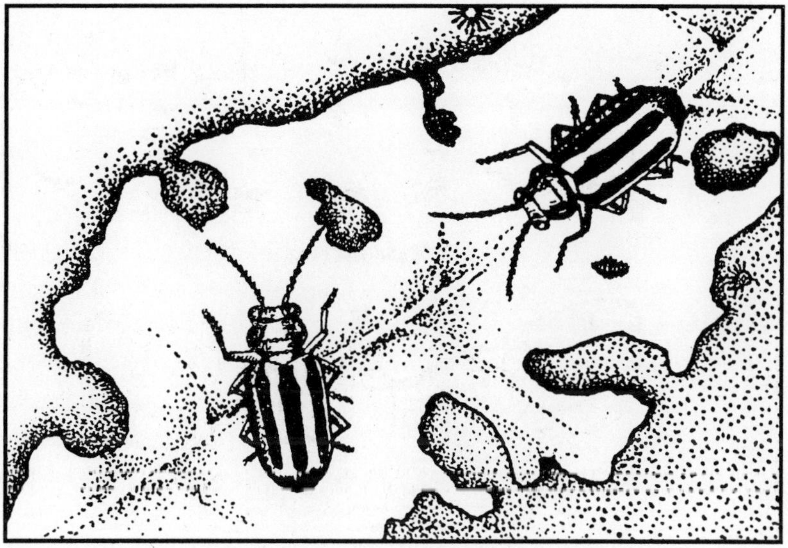

Figure 8.1: Two beetles of *Agasicles hygrophila* S. & V. feeding upon a leaf of alligator weed (diagrammatic)

Figure 8.2: *Neochetina eichhorniae* Warner. It is employed as a bioagent for the control of waterhyacinth.

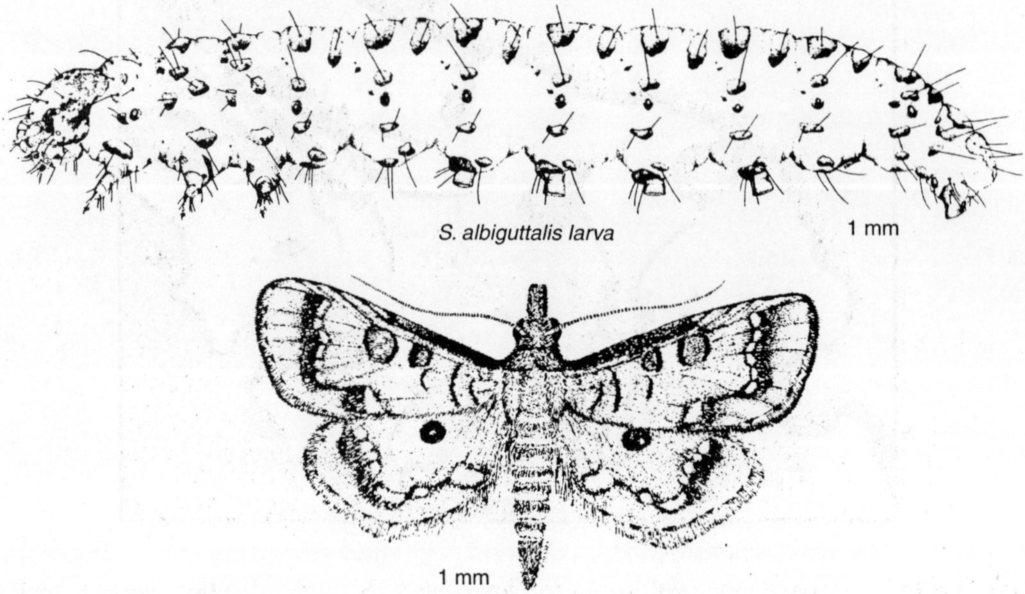

Figure 8.3: *Sameodes albiguttalis* Warren. It is employed for the bio-control of waterhyacinth. The sketch shows a larva (above) and a male adult (below).

The Commonwealth Institute of Biological Control, Trindad found three very promising arthropod bioagents against *Salvinia molesta* in South America. These insects were: (1) an aquatic grasshopper *Paulinia acuminata* De Geer, (2) a pyralid borer, *Samea multiplicalis* Guenee, and (3) a weevil *Cyrtobagous singlularis* Hulst (now *C. salviniae* Sands) (Fig. 8.4). These insects did not damage the rice and other crop plants.*

A moth *Paraponyx diminutalis* Snellen (= *Nymphula* spp.) is reported very destructive to hydrilla shoots. But fish are found to predate upon it which limits its utility.

Of late, other insect bioagents reported successful against hydrilla are: *Hydrellia pakistanae* Deonier (leaf mining fly) and *Bagous* spp. (stem boring and tuber boring weevils). These insects also attack *Potamogeton* spp. The tuber eating *Bagous offinis* operates only when the hydrilla infested waterbody has dried up.

Note: *1. *In cyclic balance, initially the population of the bioagent increases while that of the weed decreases, followed by a reverse cycle. The process keeps reversing with neither component eliminated at any stage.*

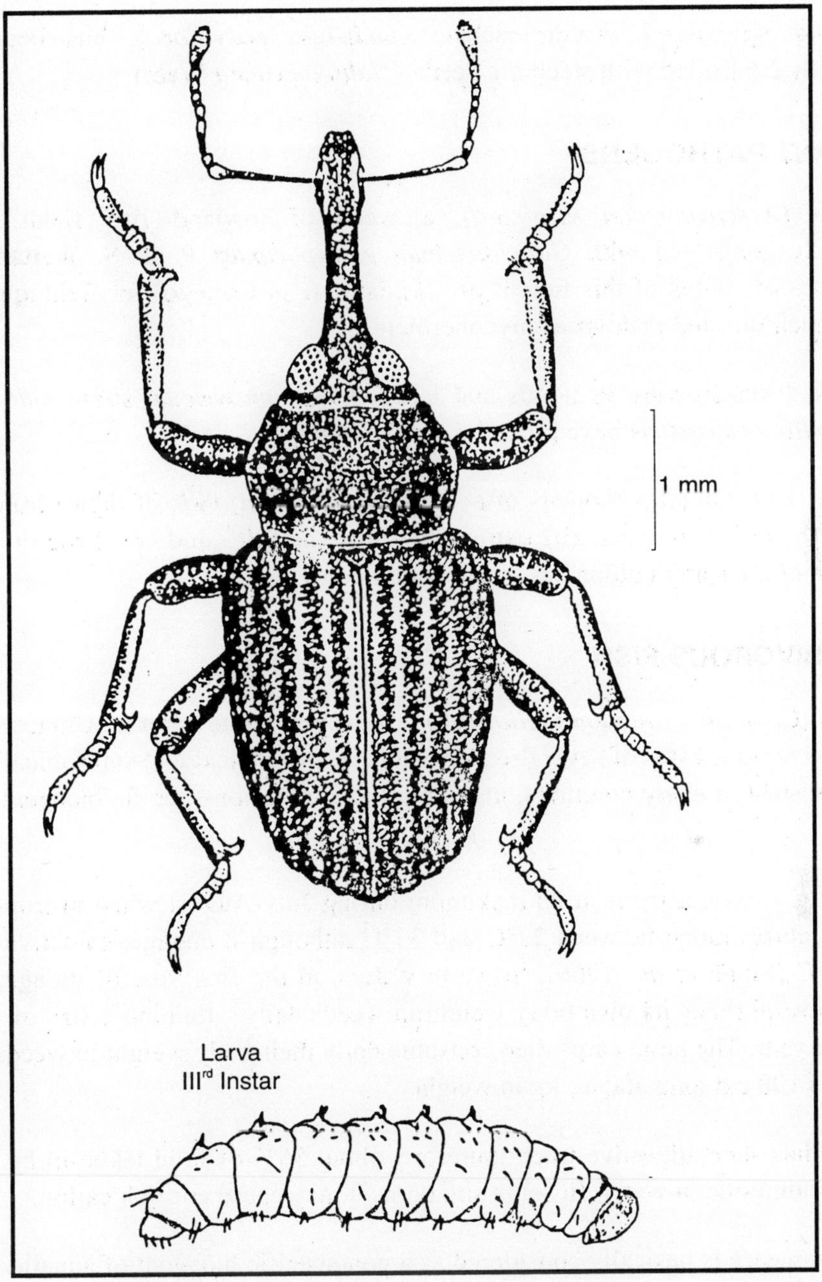

Figure 8.4 : *Cyrtobagous salviniae* Sands. It has been used as an efficient bioagent of *Salvinia molesta* in Kerala. The sketch shows a female adult (above) and a III instar larva (below).

A lowland rice weed, waterpurselane (*Ludwigia parviflora*), has been reported successfully controlled with steelblue beetle (*Haltica cyanea* Web.).

8.2: PLANT PATHOGENS

Jointvetch (*Aeschynomena virginiea*), a weed of lowland rice fields, has been successfully controlled with *Colletotrichum gloesporiodes* P. & S. Wettable powder preparations of spores of this fungus are available (*e.g.* Collego) for field application in the jointvetch infested paddies as mycoherbicides.

Against hydrilla growing in ponds and lakes, *Fusarium roseum* strain *culonorum* and *Mycdeptodiscus terrestris* have proved potential bioagents.

Biocontrol with fungal pathogens often suffers due to: (a) lack of their virulence under severe climatic fluctuations, (b) extremely short shelflife, and (c) difficulty in proper calibration of the spray equipment inside waterbodies.

8.3 HERBIVOROUS FISH

Grasscarp (*Ctenopharyngodon idella* Val.), also called white amur or Chinese carp (Fig. 8.5), is a voracious eater of several water plants in the tropical and subtropical areas. It is now established in many countries, including India, as a non-specific bioagent of aquatic weeds.*

The appetite of grasscarp is at its maximum during July-August when in tropical waters the temperatures range between 27°C and 33°C although it engages in active feeding at above 16°C (Singh *et al.*, 1966). In warm waters, in the first year of its age, grasscarp may eat several times its own body weight in weeds daily, attaining a size of 1 kg at the end of one year. The large carps then consume daily their body weight in weeds and grow as much as 100 cm long and 32 kg in weight.

Grasscarp has short digestive tract, therefore, about 35% of food taken-in by the fish is excreted undigested in water. In some situations it may cause eutrophication.

Though grasscarp is basically considered as a non-specific bioagent of aquatic weeds, yet actually it has also its strong food preferences. Its food preferences studies revealed that advanced fry and early fingerling stages of the carp preferred zooplanktons, but they also accepted tender plants of duckweeds. The advanced fingerlings and adults grazed on a

---

* *Carps are scaled fishes belonging to the family Cyprindiae. This family includes most of the cultivated fish, besides the less frequently cultivated exotic species, including grasscarp.*

number of submersed weeds of which *Hydrilla verticillata, Potamogeton nodosus*, and *Elodea nitella* were most preferred, followed by, in decreasing preferential order, *Egeria densa, P. crispus, Ludwigia peploides, P. pectinatus, Chara flexis, Eleocharis acicularis, Myriophyllum aquaticum,* and *M. spicatum. Ceratophyllum demursum* is only uprooted by grasscarp but not eaten. The carp is also found to feed upon certain filamentous algae like *Cladophora, Spirogyra,* and *Pithophora* spp.,. Notorious weeds like *Eichhornia crassipes, Salvinia molesta*, and *Pistia stratiotes* are not consumed by grasscarp beyond their occasional nibbling. Emergent vegetation likewise is not tackled by it. In the event of shortage of preferred food, it consumes the overhanging bank grasses for its survival. It may be pointed out here that when grasscarp has consumed all palatable plant species present, the less preferred submersed weeds may take over (Mc Knight and Hepp, 1995). To avoid such weed species shifts it is suggested that sequential harvesting and restocking of the carp at suitable intervals should be adopted in a manner that the preferred weed food is never completely eliminated, and a balance is maintained at sufficiently low level to permit intended economic uses of the waterbody.

It is interesting to note that while grasscarp is native to cold rivers of China and Siberia, it is successfully employed as a bioagent in the tropics and subtropics. In cold waters grasscarp exhibits limited feeding and body development but here it reproduces freely. The reverse is noted in warm waterbodies where it feeds heavily and grows fast but it seldom breeds there. Both these features of grasscarp in warm waters are considered highly desirable for its use as a bioagent. Natural, free reproduction of grasscarp in warm waters would have disqualified it as a bioagent for then it could compete with the native, better quality fish and water fowl populations. In its fry stage, grasscarp is non-vegetarian, capable of consuming eggs or fry of native, edible fish.

Despite nature's safeguard against breeding of grasscarp in tropical and subtropical waters, some scientists in the west did not accept it as adequately fool proof. This led to the development of sterile hybrids of grasscarp by crossing *C. idealla* females with male individuals of *Aristichthys nobilis* Rich (bighead carp). But the hybrid is found slower in its feeding habit than grasscarp (Osborne, 1982). In general, therefore, higher stocking rates were required for the hybrids than for the grasscarp (Shiremen, 1982). In view of such limitations of the hybrids, grasscarp has now been accepted as a practical, non-specific bioagent of submersed aquatic weeds, particularly in the flowing waters and large lakes, in many countries.

Plate 8.1 : A field trial in progress in a distributary at Kota under UNDP to evaluate the effectiveness of Chinese carp in controlling submersed vegetation. Note the weedfree water in the rear where the carp was released and the weedy, untreated water in the front. The two portions are separated by a suitable screen. (Photo-author).

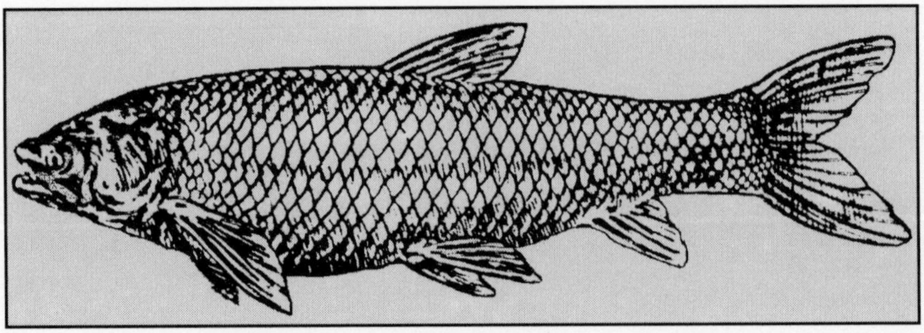

Figure 8.5: *Ctenopharyngodon idella* Val. (Chinese carp). It is an almost non-specific bioagent used for the control of submersed weeds.

For its commercial use as a bioagent, grasscarp is bred artificially using hormone injections. It matures in 2 years but the males may be oozing in 1 year. The brood fish may attain a body weight of 2-4 kg in 2 years; the males weigh less than females of the same age (Chaudhuri and Singh, 1984). In India, the grasscarp fingerlings are produced by the Central Inland Fisheries Research Institute, Barrackpore (WB). It is best to stock initially carp of 100 g each (about 10 cm long), although 150-200 g (15-20 cm long) ones may be preferred. They will be less susceptible to predators and will survive the winter well. In large waterbodies release of even adult carp may become necessary where fingerlings are likely to fall prey to some aquatic predators.

Good weed control is initiated by 100 g grasscarp at stocking rates of about 150 kg (= 1500 fingerlings) per ha water. For 300 g fish the stocking rate of 250 kg (=1250 fingerlings) per ha is found appropriate (Van Zon, 1984). But in combination with a physical or chemical weed control programme, the stocking density of fingerlings would be much less. The fingerlings gain weight quickly and after a season or two, 25% of other original grasscarp stock can be harvested regularly for food and be replaced with the younger individuals. The grasscarp of 14 kg or more body weight lose their appetite for aquatic vegetation. These should then be harvested for market without delay.

When grasscarp have cleared a waterbody of its weeds and if these are still in their active feeding stage, these should be transferred to some other weedy waterbodies for "grazing". The grasscarp cannot survive even short starvation periods. For this transfer a simple technique has been developed. It consists in first baiting the carp with hydrilla or lemna plants into some confined water pockets near the shore. Here these are sedated by pouring a mixture of rotenone* and potassium permanganate in water. The carp can then be easily picked-up in water containers and transferred to the new waterbody.

---

* *Note: Rotenone is obtained from roots of Derris elliptica. It is sold under several trade names, e.g. Fish Tox and Nox-Fish.*

Considerable caution is necessary in importing grasscarp from an alien country. Each lot of carp imported could carry new fish pests and diseases. Therefore, these should be quarantined properly before permitting their entry in any country. Also, once the fingerlings are released in weedy waterbodies there should be some regulation on fishing these, both in respect of the waiting interval and the minimal size of fish at harvest. Usually carp should reach a minimum length of 30-40 cm at the first fishing time. One should avoid removal of the grasscarp before these had chance to control the weeds, effectively.

Besides grasscarp, there are three other fish of value as bioagents, *viz.*, silver carp (*Hypophthalmichthys molitrix* Val.), common carp (*Cyprinus carpio*) and tilapia (*Tilapia nilotica* L. and *T. mossambica* Peters). These are found great consumers of weedy algae. *T. mossambica* is particularly useful for algal control in paddy fields since it does not damage the rice plants.

8.4: MANATEE (*Trichechus manatus* L.)

Manatee, also called seacow (Fig. 8.6), is totally herbivorous, aquatic, large (up to 4 m long with 500 kg body weight), warm blooded mammal. These animals consume huge quantities of aquatic vegetation at a time. But unfortunately, the number of manatees in the world left presently, after their selfish hunting by man for their meat, are so little that these are now becoming fast extinct. Nowhere in the world the manatee populations are large enough to support any significant aquatic weed control programme. Artificial breeding of this animal in captivity or semi-captivity has not been successful so far but the efforts must continue at the international level. Perhaps, confinement and nutritional deficiencies are the major reasons for this behaviour. Now the only hope to enlarge the few mild populations of manatee known to exist in the tropical areas, is their world-wide protection in their natural conditions by law. Hunting manatees is wasteful, undesirable, and irresponsible. The International Manatee Research Station at Guyana has taken some steps in this direction. Let us hope someday we will have enough manatees to solve our several aquatic weed problems.

8.5: OTHER ANIMALS

In China, ducks have been employed to destroy succulent weed seedlings in lowland paddies as a supplementary weed control method. But the practice requires thousands of ducks at a time and considerable man power to herd them. The ducks can be marketed after they are about 1 kg each.

Large rodents like *Capybara*, and *Nutria* are found to feed upon waterhyacinth. But this can be considered only an academic observation since no one will take to multiply rodents as bioagents.

Figure 8.6: *Trichechus manatus* L. (manatee). It is an aquatic mammal weighing 250-1000 Kg. It can be used as a very good non-specific bioagent of several aquatic weeds but presently it is facing extinction at the hands of man.

8.6: COMPETITIVE PLANTS

About six decades ago it was observed that when *Azolla* was grown with transplanted rice, only one hand-weeding was needed instead of the usual two or three. Recent scientific investigations by Jaiya and Moody (1984) confirmed the role of *A. pinnata* R. Br. in suppressing 69-100% growth of *Monochoria vaginalis*, *Echinochloa glabrescens*, *Cyperus difformis,* and *Paspalum* spp., when the fern was established before emergence of the weeds.

Weed scientists have since been looking for aquatic plants that would quickly cover the hydrosoil and prevent weed establishment there without hindering the waterflow. In this respect the potential of *Eleocharis acicularis* (L). R.B.D. (slender spikerush) to oust many troublesome rooted, aquatic weeds from canal bottoms was first demonstrated in California in 1966 (Yeo, 1980). *E. acicularis* in itself is a short plant, incapable of hindering water flow in canals, although it can do so in shallow irrigation ditches. The plant spreads laterally up to 2 m through its rhizomes and forms tufts that will not allow other aquatic weeds to easily establish. Slender spikerush propagates also with tubercles and achenes. *E. acicularis* survives low light intensities available at depths down 3 m. Also, it tolerates common aquatic herbicides used in canals, including acrolein. Thus, it may warrant its inclusion as a bioagent in integrated biological and chemical aquatic weed control programmes in canals. In California this plant has been successfully exploited to compete with *Potamogeton, Elodea*, and *Zennichellia* spp.

Two other species of spikerush, namely *E. coloradensis* and *E. parvula* were reported by Darr (1984) to be equally effective in displacing certain submersed aquatic weeds from canal bottoms.

Other competitive plants, similar in action to *E. acicularis*, are *Isoetes bolanderi* and *Marsilea mucronate*. Also, persistent waterblooms have been reported to effectively shade out *Potamogeton pectinatus* in California. The bloom was encouraged by adequate fertilisation of the lake.

In India, the regrowth of *Typha* spp. (cattails) in ditches was checked by planting *Brachiaria mutica* L. (paragrass). The grass was found highly competitive to cattails and it provided valuable forage for livestock (Mehta and Sharma, 1975). Likewise by encouraging the growth of broadleaf weeds in ditches, invasion of *Sorghum halepense* could be reduced. In Yugoslavia, mosses were reported to successfully compete with several grass weeds on the river banks.

Of late, **allelopathic approach** to the control of aquatic weeds has been investigated. Kathiresan (1999) claimed that dry powder of *Parthenium hysterophorus*, *Coleus amboinicus*, and *Cuscuta reflexa* plants applied evenly in a waterhyacinth-infested waterbody could destroy the weed in 15 days. The effective application rates of powder of these plants were: 12.5-30.0 g/l of water present in the weedy waterbody for the first two plants and 30.0 g/l of water in case of *C. reflexa*. The spray of either plant powder on the hyacinth plants was not found effective in killing the weed.

Although the addition of certain allelopathic plant powder to waterbodies to kill waterhyacinth seems an attractive proposition, yet the cost of procuring huge quantities of the allelo-plant powder needs be worked out. Perhaps, it may be more worthwhile to attempt to extract allelo-chemicals from these plants for their eventual commercial synthesis and use as herbicides against waterhyacinth than their direct use.

SELECTED BIBLIOGRAPHY

1. Allitop, W.H.L. 1960. The manatee: ecology and use for weed control. Nature (London) 188 (4782). 762.

2. Bennett, F.D. 1966. Investigations on insects attacking the aquatic ferns, *Salvinia* spp. in Trindad and N-S America. Proc. South. Weed Conf. 19: 497-504.

3. Bonar, S.A. *et al.* 1996. Management of aquatic plants in Washington state using grasscarp : effects on aquatic plants, water aqulity, and public satisfaction 1990_1995. Inland Fish Dev. Res. Washington, Dept, Fish & Wilde. Olympia, WA, 28 p.

4. Bosley, R.W. 1975. White amur, the wonder fish, solves the water weed problems. Amer. Nurseryman 141 (9): 9, 83.

5. Brabben, t. 1996. Developments in aquatic weed control. GRID, IPTRID Network Magazine No. 8, 3.

6. Buckingham, G.R. 1996. Biological control of alligatorweed. Castanea 61 (3): 232-243.

7. Cassant, J.R. and W.E. Caton. 1983. Feeding behaviour of yearling and older hybrid grasscarp. J. Fish Biology 22 (1): 35-41.

8. Center, T.D. 1984. Dispersal and variation in infestation intensities of waterhyacinth moth *Sameodes albiguttalis* (Lepidoptera: Pyralidae) populations in Peninsular Florida. Environ. Ento 13 (2): 482-491.

9. Center, T.D. 1994. Biological control of weeds: Waterhyacinth and waterlettuce. (*IN*) Pest Management in the subtropics (*ed.* by Rosen, D. *et al.*). Intercept Ltd., USA.

10. Center, T.D. *et al.* 1982. Descriptions of *Sameodes albiguttalis* life stages with key to Lepidoptera larvae on waterhyacinth. Annals of Ento. Soc. Amer. 75 (4): 471-479.

11. Center, T.D. *et al.* 1982. Control of waterhyacinth with *Neochetina eichhorniae* and a growth retardant. Weed Sci. 30 (5): 453-457.

12. Charudattan, R. 1982. Biological control of weeds with plant pathogens. Jhon Willey and Sons.

13. Charudattan, R. 1998. Plant pathogens for biological control of waterhyacinth-A status report. Abstr. First Global Working Group Meet for the Biol. and Integr. Control of Waterhyacinth. IOBC, Harare, Zimbabwe: 9

14. Chaudhuri, H. and S.B. Singh 1984. Induced breeding of carp. Ind. Council of Agri. Res., New Delhi (India) Publication, 82 p.

15. Couch, R.W. 1990. Managing algae growth with grasscarp and algaecides. Weed Abst. 42 (6): 2587.

16. Darr, S. 1984. Using plant competition to control aquatic weeds. IPM Practioner 6 (1): 2-3.

17. Deloach, C.J. 1982. Natural enemies of *Neochetina bruchi* and *N. eichhorniae*. Ann. Entomol. Soc. America 75 (2): 115-118.

18. Edwards, D.J. 1974. Weed preference and growth of young grasscarp in New Zealand. New Zealand J. Marine and Freshwater Res. 8 (2): 341-350.

19. Elakovich, S.D. 1989. Allelopathic aquatic plants for aquatic weed management. Biologia Plantarum 31 (6): 479-486.

20. Forno, I.W. *et al.* 1983. Distribution, biology, and host specificity of *Cyrtobagous singularis* Hulst for the biological control of *Salvinia molesta*. Bull. Ent. Res. 73, 85-89.

21. Godfrey, K.K. and L.W.J. Anderson. 1994. Feeding by *Bagous affinis* (Coleoptera) inhibits germination of hydrilla tubers. Florida Entomologists 77 (4): 480-488.

22. Gupta, O.P. 1993. Aquatic weed management. Proc. Int. Symp. Integrated Weed Management For Sustainable Agriculture. Ind. Soc. Weed Sci., HAU, Hisar (India): 399-406 pp.

23. Hanlon, S.G. 1999. Evaluation of macrophyte control in 41 Florida lakes using triploid grasscarp at different stocking rates. Dissertation Abs.Int. Ann Arbor, Michigan, USA, 113 P.

24. Harley, K.L.S. 1981. Successful biological control of the floating weed *Salvinia*. Nature 294 (5): 78-80.

25. Harley, K.L.S. 1984. Biological control of waterlettuce. J. Aquat. Plant Management 22: 101-102.

26. Hickling, C.F. 1967. On the biology of herbivorous fish--the whiteamur or grasscarp. Proc. Royal Soc. Edin. (B) 70 Part, 1 (4): 32-81.

27. Holtikamp, R. 1996. Biological control of waterhyacinth. Agnote-NSW Agriculture 2/092, 2p.

28. Janiya, J.D. and K. Moody. 1984. Use of azolla to suppress weeds in transplanted rice. Trop. Pest. Manage. 30 (1): 1-6.

29. Joy, P.J. *et al.* 1985. Successful biological control of the floating weed *Salvinia molesta* Mitchell using the weevil *Cyrtobagous salviniae* Sands in Kerala (India). Proc. Tenth Asian-Pacific Weed Science Society, 622-626.

30. Julien, M. 1997. Biological Control of Weeds--Theory and Practical Application, M. Mulien and G. White (*Ed*), ACIAR Monograph No. 49, 1-8 pp.

31. Julien. M.H. and M.J. Storrs. 1996. Integrating biological and herbicidal controls to manage *Salvinia* in Australia. Weed Abst. 45 (12), 4996.

32. Kanchan, S.D. and Jayachandra, 1979. Allelopathic effects of *Parthenium hysterophorus* exudation of inhibitors through roots. Plant and Soil Science 53: 27-35.

33. Kathiresan, R.M. 1999. Use of bioagents in weed management. Souvenir 8th Biennial Conference of the Ind. Soc. Weed Sci. held 5th-7th Feb. 99 at Varanasi (India): 41-46.

34. Kulshreshtha, M. and B. Gopal. 1983. Allelopathic influences of *Hydrilla verticiallata* on the destruction of *Ceratophyllum* species. Aquat. Bot. 16 (2): 207-209.

35. Maddox, D.M. *et al.* 1971. Insects to control alligatorweed--an invader of aquatic ecosystem in the US. Bio. Sci. 21 (19): 985-991.

36. Marois, J.J. 1994. Integration of biological control into IPM system for aquatic weeds. Weed Abst. 45 (4): 1644.

37. Mc. Knight, S.K. and G.R. Hepp. 1995. Potential effect of grasscarp herbivory on waterfowl food. J. Wildlife Management 59 (4): 720-727.

38. Mehta, I. and D.S. Boonlia. 1980. Weed out weed with a weed. Intensive Agri. 18 (6): 23-24.

39. Mehta, I. and R.K. Sharma. 1972. Control of aquatic weeds by the white amur in Rajasthan. Hyacinth Contr. J. 10: 16-19.

40. Netherland, M.D. and J.F. Shearer. 1996. Integrated use of fluridone and a fungal pathogen for control of hydrilla. J. Aq. Plant Manage. 34, 4-8.

41. Orapa, W. and C. Atip. 1996. Update on biological control of waterhyacinth in Papua New Guinea. Didimag News Letter 28 (9-12): 28-30.

42. Osborne, J.A. 1982. Herbicice v/s grasscarp. Aquatics 4 (2): 13.

43. Pandey, D.K. 1994. Inhibition of salvinia (*S. molesta*) by parthenium. J. Chem. Eco. 20 (12), 3111-3122.

44. Pandey, D.K. *et al.* 1993. Inhibitory effect of parthenium residues on growth of waterhyacinth. J. Chem. Eco. 19 (11): 2561-2.

45. Prowse, G.A. 1971. Experimental criteria for studying carp feeding in relation to weed control. Progressive Fish Culturist 33 (3): 128-131.

46. Room, P.M. *et al.* 1981. Biocontrol of *Salvinia* in Australia. Nature 294.

47. Sands, D.P.A. 1983. Identity of *Cyrtobagous* sp. (Coleoptera: Curculionidae) introduced into Australia for biological control of salvinia. J. Aquat. Ento. Soc. 22: 200.

48. Sand, D.P.A. *et al.* 1983. The feeding characteristics and development of larvae of a salvinia weevil, *Cyrtobagous* sp. Ento. Expt. and Appl. 34: 291-296.

49. Shabana, Y.M. *et al.* 1995. Evalution of *Alternaria eichhorniae* as a bioherbicide for waterhyacinth. Biologial Control 5 (2): 136-144.

50. Shearer, J.F. 1998. Biological control of hydrilla using an endemic fungal pathogen. J. Auqat. Plant Manage. 36: 54-56.

51. Shireman, J.V. 1982. Cost analysis of aquatic weed control; fish versus chemicals in a Florida lake. Progr. Fish Culturist 44 (4): 199-200.

52. Singh, S.B. *et al.* 1966. Observations on the efficacy of grasscarp in controlling and utilizing aquatic weeds in ponds in India. Proc. Indo-Pacific Fish Council 12 (2): 220.

53. Sivakumar, C. and R.M. Kathiresan. 1997. Weed management in rice-fish farming system. Abstr. Biennial Conf. Ind. Soc. Weed Sci., Punjab Agri. Univ.,: 73.

54. Stark, J.D. 1983. Life cycle and behaviour of *Neochetina eichhorniae* in Lousiana. Environ. Ent. 12: 147-150.

55. Stewart, C.A. *et al.* 1995. The potential geographical distribution of alligator weed and a biological control agent *Agasicles hygrophila* in New Zealand. Weed Abst. 45 (11): 4554.

56. Stott, B. and L.D. Orr. 1970. Estimating the amount of aquatic weeds consumed by grasscarp. Progressive Fish Culturist 32 (1): 51-54.

57. Sutton, D.L. 1975. A handhoe with fins. Weeds Today 6 (1): 14-15.

58. Sutton, D.L. 1984. Utilisation of hydrilla by the whiteamur. Hyacinth Control J. 12 (May): 66-70.

59. Thielen, R. *et al.* 1994. Importation, release, and establishment of *Neochetina* spp. for the control of waterhyacinth in Benen (W. Africa). Entomophaga 39 (2): 179-188.

60. Van Zon, J.C.J. 1977. Grasscarp (*Ctenopharyngodon idella* Val.) in Europe. Aquat. Bot. 3: 143-155.

61. Van Zon, J.C.J. 1984. Economic weed control with grasscarp. Trop. Pest. Manage. 30 (2): 179-185.

62. Visalakshy, P.N.G. and K.P. Jayanth. 1996. Effect of silt coverage of waterhyacinth roots on pupation of *Neochetina eichhorniae* and *N. bruchi*. Biocontrol Sci. and Technol. 6 (1): 11-13.

63. Yeo, R.R. 1980. Life history and ecology of dwarf spikerush (*Eleocharis coloradonensis*). Weed Sci. 28 (3): 363-372.

Chapter 9

AQUATIC PLANT MANAGEMENT IN PISCICULTURE

9.1: ESSENTIALITY OF AQUATIC PLANTS IN PISCICULTURE

Directly or indirectly, fish, waterfowl, and other large aquatic animals depend for their food on aquatic plant life. Phytoplanktons form the primary organisms in the food chain of all aquatic ecosystems. Zooplanktons and benthic micro-invertebrates, which form secondary food in the aquatic food chains, use all kind of organic matter formed from the decay of plants as their important energy sources. Besides these primary and secondary organisms, herbivorous fish utilise many aquatic herbs directly as their sole food. Too, the photosynthetic releases of oxygen by aquatic plants is extremely important for meeting the respiration needs of aquatic fauna. An aquatic plant like *Hydrilla* helps fisheries by reducing the turbidity of water. The shoreline aquatic plants, particularly the emersed types, offer protective breeding, nesting, and play sites to fish (and ducks)*. Wildrice and wildmillet furnish food to waterfowl. The low growing, bank grasses improve economic value of fishery waterbodies by preventing bank erosion and silting.

---

* *Note:*

1. *Some common shelter aquatic plants for fish are: Alisma, Brasenia, Ceratophyllum, Lemna, Marsilea, Nelumbo, Nuphar, Nymphoides, and Wolffia spp.*

2. *Some common aquatic food plants of fish are: Cabomba, Chara, Cladophora, Heteranthera, Nasturium, Najas, and Wolffia spp.*

9.2: DAMAGE TO PISCICULTURE BY OVERGROWTH OF PLANTS

Despite the great importance of varied aquatic flora in fishery waters (and duck marshlands), plant growth that does not justify the space it occupies in and around these waterbodies, is definitely a big negative factor in economic production of fish and other aquatic sport animals in many ways (Chapter-1).

In India, and many other countries of the world, fishery waters abound badly in weedy vegetation, requiring their restriction to direct the energy flow in the ecosystem to the advantage of the edible or sport aquatic fauna. Aquatic ecosystem models to quantify the effects of macrophytes and algae on the fish and other economic fauna need be developed (Canfield, *et al.* 1983). Such studies will enable the owners to decide how much vegetation should be brought under control in a given fishery waterbody (or duck marsh) so that additional energy would flow towards the production of primary and secondary aquatic consumer organisms.

According to the estimates of Philipose (1968), in India about 40% of the cultivated fishery waters, amounting to approximately 800,000 ha, suffered from excessive growth of aquatic vegetation. The problem was particularly severe in the eastern states where up to 70% of fish water area was found infested with aquatic weeds (Philipose *et al.*, 1970). Also, several low lying lands used as seasonal fish grounds had been converted by weeds into unproductive swamps. All these estimates were made about three decades back. Today the situation is much worse.

In fishery waters, of the several weed species, *Eichhornia crassipes, Hydrilla verticillata*, and *Potamogeton* spp. are particularly obnoxious. Besides, the waterbloom forming algae (chiefly *Microcystis* and *Anabaena*) and scum algae (chiefly *Pithophora* and *Spirogyra*) are problematic in pisciculture.

9.3: WEED CONTROL OPTIONS IN FISH PONDS

Weed control in fish culture begins with preventive measures at the designing stage of new waterbodies and it continues into their periodic, physical clearances and decreasingly intensive but constant efforts through the chemical and biological resources. Also, when possible, renovation of badly weedy fish ponds is often considered desirable.

9.3.1: Preventive Measures

The first step in aquatic weed management in fish ponds is their proper **location and design.** In principle, extensive, shallow lands should be avoided for the construction of fish ponds since these areas are likely to receive frequent drainage, enriched with plant nutrients and pollutants. In fish ponds located on farms the site should be so chosen as to minimise run off from the fertilized fields. A site where the ponds could be drained and refilled at will is ideal for fishery development from the point of view of keeping the noxious vegetation under check. It will also help in the renovation of fish ponds when these get aged. Small farm ponds are renovated once every 3-4 years during summers. The drainage of ponds is very useful in allowing the bottom muck to decompose rapidly and, thus, prevent development of high B.O.D. during the active growth period of fish.

Fish ponds should have little, if any, water areas of 1 m or less depth during any part of the year. The submersed weeds establish initially in shallow waters and then they gradually invade the deeper water areas of the ponds and lakes. The shallow shorelines of 1 m or less depth should be deepened and the dug mud should be used to fill marshy spots around the pond.

9.3.2: Control Measures

(a) **Physical methods:** In India, in small fish waterbodies various traditional manual methods, including pulling and netting are employed. For removing the submersed weeds, log weeders with spikes and barbed wires; hand-drawn bottom rakes; bamboo poles with basal cross attachment or boom forks having toothed prongs; and long-handled forks and sickles are often successfully used. In West Bengal and N-E states some fish farmers have used launches having long sickles mounted at the bow in the form of a `V' with reciprocal action for cutting emergent weeds like *Colocasia* and *Cyperus* spp. In Assam, thick infestations of *E. crassipes* (waterhyacinth) are dragged to the margins by power winches, and later pulled out with mechanical lifts. In Orrisa, power weed-cutters are employed to clear thick infestations of rooted submersed weeds (Philipose, 1968).

Sometimes, frequent physical removal of weeds from fishery waters may deplete it of its mineral nutrients needed for the growth of phytoplanktons. Such nutrient losses should be compensated by addition of fertilizers to the fish ponds. When physical methods of weed control are supplemented with herbicides or bioagents, the rate of nutrition depletion from a waterbody is reduced. It all depends upon how nutrient rich the fishery water was initially and what were other sources of supplying nutrients to the fish ponds.

Several alternative **shading** materials to cut light to aquatic weeds, particularly to the submersed types, have been tried in an attempt to destroy them. Fibreglass screens coated with black PVC proved ineffective since these still absorbed 30% of light (Engel, 1984). The cover of a lightweight, gas permeable, opaque, spun-bonded, polypropylene textile fabric controlled submersed vegetation in 6-12 weeks, depending upon the weed species. Covering areas of water with black polyethylene or PVC gave excellent control of submersed weeds and waterblooms but it was practicable only on fingerling tanks and like small ponds.

(b) Biological methods: Biological control of excessive aquatic plants with grasscarp is important in fisheries. Grasscarp is a herbivore and it is almost incapable of breeding in tropical and subtropical waters. Besides, it has a large appetite for several submersed weeds. All these features make carp a non-competitive, excellent bioagent in fisheries (Von Zon, 1984). Also, since the grasscarp is easily identified from other fish, it is easy to trap and transfer it to new waterbodies for weed control (Chapter 8).

(c) Herbicidal control: Some aquatic weed problems can be usefully tackled with herbicides. An ideal herbicide for use in pisciculture should possess the following redeeming features: -

(1) *It should be safe to fish and fish food to the extent that the growth rates of fish in the herbicide-treated waters are not less than those expected in the otherwise weedfree waters.*

(2) *The herbicide should get metabolised in the body of the fish rapidly to levels safe for humans.*

However, in practice, some damage caused by a herbicide to fish should be weighed against the increased fish production due to effective weed control before it is rejected. In fish ponds covered with weeds, no meaningful fish production can be expected any way.

The following herbicides are commonly permitted in fisheries: -

(i) ***Copper compounds***: Copper sulphate pentahydrate (CSP) is effective in controlling algae in fisheries at 0.2-2.0 ppmw. But now chelated Cu formulations are preferred over CSP because of their greater margin of safety to fish, as well as for their greater potency in hard or turbid waters. Cutrine, Cutrine plus, Komeen, and K-Lox are some common chelated Cu formulations employed to control algae and hydrilla in fisheries. Despite their safety to fish, their frequent use in the same waterbody should be avoided to guard against possible accumulation of Cu in the food chain. No doubt, when there is a choice between the scum algae and Cu algicides, one is justified in choosing the latter.

Do not apply Cu compounds in trout waters if the carbonate hardness of the water does not exceed 50 ppm.

(ii) ***Diquat***: Diquat is a versatile herbicide to treat floating weeds like *E. crassipes*, *Pistia stratiotes*, and *Lemna* sp, as well as some submersed weeds in still and slow flowing ponds, lakes, and canals. It is safe to planktonic fish food. The common rates are 1-2 kg ha^{-1}. It is believed that at the recommended rates diquat is non-toxic to fish but the *cationic surfactants included in its common formulations may sometimes harm fish.* As such, only special, surfactant-free diquat formulations are used in fisheries.

Against submersed weeds, apply diquat to the surface of water or inject below it and distribute as evenly as possible. Repeat when new infestation occurs. Diquat action is reduced in muddy waters due to its quick adsorption on clay particles. It is often applied with Cu compounds.

Diquat may be irritating to the skin. It is available as aqueous solution. The treated water can be used for irrigation.*

(iii) **S-*Traizines*:** Simazine, ametryn, and terbutryn are found safer algicides in fisheries than Cu compounds. Their effective rates are: 0.05-0.1 ppm for ametryn and terbutryn, and 0.5-1.24 ppm for simazine. Among the three triazines, terbutryn is considered safest to fish but trout are susceptible to it. The slow release granules of *s*-triazines are usually preferred in fisheries over their wettable powders. Besides algae, certain submersed weeds are also controlled with these. Now floating formulations of the three *s*-triazines are also available to maintain their high concentrations in the waterbloom zone of the water. An important advantage of using *s*-triazines as algicides is that these spare enough phytoplanktons (and all diatoms) to serve as fish food. Also, the triazines kill algae only slowly by distracting their photosynthesis. The complete kill occurs in about 16-21 days. This prevents sudden rise in the B.O.D. of the waterbody and allows new growth of zooplanktons for fish. Sometimes, triazine resistant biotypes of algae may be present.

Do not try to use atrazine instead of simazine since the fomer is highly toxic to fish.

(iv) ***2,4-D:*** A major use of 2,4-D in fisheries is to sink the waterhyacinth mats. The treatment may impart some *temporary,* unpleasant flavour to fish but it is acceptable because under the waterhyacinth cover there can be no economic production of fish at all. Moreover, when prescribed waiting periods are observed, there is little likelihood of any disflavour left in fish. Ahmed *et al.* (1982) found that dimethyl amine 2,4-D applied at 3.34 kg *a.i.* ha^{-1} to a hyacinth-infested fishery

---

Note: 1. See Chapter-7 for primary information on aquatic herbicides and algicides.

lake was degraded in water to 0.1 ppm level within 10 weeks after its application. This waiting period, in fact, is an unavoidable compulsion since it will take considerable time before the fish production is revived to levels suited for harvesting.

The slow release pellets of amine 2,4-D are usually not permitted in fishery waters because these are hydrolised only slowly in the body of the fish. But BEE of 2,4-D is considered comparatively safe to fish.

(v) ***Endothall***: The granules of Na and K endothall are freely used for treating fish ponds and lakes to control difficult weeds like *Potamogeton, Hydrilla, Ceratophyllum, Cabomba, Lemna* spp. and some algae at 0.2-0.5 ppm. Aquatic formulations (inorganic salts) are safe to fish in 100-500 ppm concentrations. Spray, inject, or apply as granular material when weeds are actively growing. Water containing dense vegetation should be treated in sections, 5-7 days apart, to prevent suffocation of the fish.

Do not use fish from endothall-treated water within three days after treatment. Endothall may be toxic to certain fish; the amine salt being more so than the acid.

The diamine salt of endothall is very potent on weeds but it is toxic to fish, therefore, its use is limited to the treatment of shorelines. In large lakes it can also be used for their localised or zonal treatment (Chapter 7).

The slow release pellets of monoamine salt of endothall are found particularly effective against hydrilla, without any serious harm to most fish.

Endothall is irritating to eyes, nose, skin, and throat.

(vi) ***Silvex*** (fenoprop): Silvex is used in fisheries against *Alternanthera, Nymphaea*, and *Nymphoides* spp., although some submersed weeds are also susceptible to it. Its effective rates are 1-4 kg ha^{-1}. It should be applied in sufficient water to wet the foliage. Apply when the weeds are actively growing but before these have formed dense mats. One application is usually sufficient. Begin the application at the water edges along the entire shoreline, continuing the operation in succeedingly smaller circles until the entire area has been covered. In larger waterbodies follow the localised application technique. Water can be used for recreation soon after the treatment. Do not apply CSP to fish pond for 3 days after or before treatment with silvex. *Some kinds of fish may be killed in the silvex treated water.* Water temperature should be above 10°C at the time of treatment but not too warm to induce its volatilisation loss.

Fishing should be permitted after the prescribed waiting period of the formulation used.

(vii) ***Sodium arsenite***: Sodium arsenite is absolutely safe to fish even at 8-10 ppm, although several submersed weeds and algae are killed at half these rates. *A major limitation of sodium arsenite is its **high mammalian toxicity.*** Therefore, it should be applied only under expert supervision.

(viii) ***Diuron***: It is used as an algicide in similar fashion as *s*-triazines. Its effective rates are 0.02 ppm against waterbloom and 0.2-0.4 ppm against macroalgae.

In general, zonal application of herbicides to only a portion of a large waterbody at a time in variance with its one time, full area treatment permits the use of even those short persistence herbicides that are usually considered toxic to fish since after the treatment the fish can move to the uncontaminated areas of the waterbody. The zonal treatment also avoids sudden rise in B.O.D. of the waterbody (Chapter-7).

9.4: RENOVATION OF DEGRADED FISH PONDS

When fish ponds and lakes are plagued with dense masses of weeds to the extent that meaningful fish culture is no longer possible in them, it is worthwhile to renovate them. Conventionally, the fish ponds are renovated by their complete drainage and manual digging of the ponds. Their bottoms are then left open for drying in sun to desiccate the underground weed propagules. But this practice often proves unsatisfactory. Soon after the ponds are re-filled with water, the weeds stage a come back. Treatment of the exposed bottoms of fish ponds with either dichlobenil (5-10 kg ha^{-1}), simazine (10-20 kg ha^{-1}), or fenac (15-20 kg ha^{-1}) can hold back the weeds for considerable time.

Fish ponds can also be renovated without draining them by injecting herbicides like ammonia, acrolein, and dichlobenil in water. Of these, in India liquor ammonia (9-10 ppm) is found most economical to use (Phillipose, 1963). After the weeds are disintegrated, ammonia serves as a fertilizer to rebuild the planktonic growth in fish ponds. Also, ammonia will cause total kill of predatory fish and fish parasites. Once a pond is cleared-off of all its unwanted flora and fauna, one could start afresh with new fingerlings in a weedfree environment. *Ammonia and acrolein must be applied with great operational care as already discussed in Chapter-7.*

9.5: DIRECT EFFECTS OF HERBICIDES ON FISH

The direct effects of herbicides on fish growth are measured both in laboratory tubs and field ponds by adding the test herbicide in weedfree fishery waters and measuring the gains or losses in the size and weight of the test fish. Also, the scientists watch the activity of the test fish in the treated and untreated waters. An adverse herbicide environment makes the fish restless initially and sluggish subsequently. The fish then fails to avoid a hand net. Finally, the affected fish may lose their swimming capability. This biotest of herbicides on fish is called **avoidance of net test**.

Despite limited scientific information available at this point, there is little evidence to doubt that the approved aquatic herbicides have been detrimental to the fish environment and that residues have accumulated, save **As and Cu** based herbicides used in confined waterbodies. There are reports that frequent use of copper compounds in some western countries reduced fish production because of accumulation of Cu in water. Same was found true of arsenic. Therefore, their frequent use in confined waterbodies should be avoided. The Cu and As ions absorbed by weeds are eventually recycled in water upon their decomposition. *Accumulation of As up to 4 ppm and of Cu up to 1 ppm in water would prove toxic to fish and trouts, and probably also to their food organisms* (Mullison, 1970; Lawrence, 1957). The tolerance limits of several kinds of fish to CSP varies between 0.14-2.0 ppm.

Diquat and paraquat are accumulated on mud particles for years by adsorption. But this adsorption is irreversible and so the two herbicides should not pose any residue threat to fish. Granular **fenac, dichlobenil, and 2,4-D** also have been found to persist in hydrosoil for long in confined waters (Frank and Comes, 1967), though in flowing waters, these are rapidly diluted to safe levels after an hour or so (Frank *et al.*, 1967). Dichlobenil and 2,4-D at their common rates, however, were found safe to fish and fish food even in confined waters although waiting periods are prescribed for fishing (Taylor, 1966 and Coper *et al.*, 1969). This waiting period was found short in case of 2,4-D, but up to 2 months for dichlobenil. Yeo (1970) found that on an average 71% of the applied endothall (Na and K salts) dissipated from confined waters in 12 days. Hiltibran (1962) reported endothall persisted in water only for 4 days, with no residues in fish.

Simazine (but not atrazine) is very safe to fish and its reproductive products. However, sometimes the destruction of phytoplanktons by its application may reduce fish growth, temporarily. Even CSP, 2,4-D, silvex, and diquat may adversely act on some component of food chain of fish, and thus reduce fish yields for short duration (Hiltibran, 1967). A transient reduction in planktonic growth for about 2 weeks was caused by 1.2 ppm silvex in fish ponds.

Diuron was found to accumulate in fish tissues. Also, its decomposition product dichloraniline was found toxic to fish. But at the rates needed for algae control, such detrimental effects of diuron have not been recorded.

Table 9.1: LD_{50} Of Certain Herbicides Present In Water For Fish (After CSIRO, Australia, 1984, Technical Report No.2)

Herbicide	*LD_{50} (ppm in water)*	*Average concentration required for weed or alga control (ppm)*	*Safety rating*
Dichlone	0.05	0.5	Not safe
Diquat	12.0	Variable	Safe
Endothall amine	2.0	2.0	Not safe
Diuron	3.0	0.05	Safe
Hexazinone	300.0	0.5	Very safe
Simazine	5.0	1.0	Safe
Ametryn	5.0	1.0	Safe
Terbutryn	3.0	0.5	Safe
Chelated Cu	100.0	0.2	Very safe
Copper sulphate	i. 1.0 in soft water	0.2	Safe
	ii. 7.0 in hard water	2.0	Not safe

Comment: *A high LD_{50} value of any herbicide alone is not sufficient to permit its indiscriminate use in fisheries. One must consider at the same time rapidity of deoxygenation created after its application to fishery waters, as well its effects on fish food. LD_{50} is only an initial guide to establish safety of a herbicide to fish.*

The **kind of formulation** of at least certain herbicides, notably 2,4-D, silvex, endothall, and diquat, greatly influences their toxicity to fish or fish food. Endothall (acid) and its slow release formulations are very safe to fish while dimethylalkyl amine salt of endothall as W.P. is 100 times more toxic to them. Likewise, PGBEE of 2,4-D and silvex are found much safer to fish than its other esters. Using radio active compounds it was found that 2,4-D (acid) was not absorbed from water by the gill membranes of rainbow trout, channel catfish, and bluegills. Also, when it was taken in by the fish through mouth, it was excreted through their kidney and gills. The BEE of 2,4-D and silvex are rapidly hydrolysed to their acid forms in water, as well as inside body of the fish. As such, these formulations also do not pose any hazard to the normal growth of fish. Other esters of 2,4-D and silvex are not hydrolysed rapidly and, thus, they may prove toxic to fish (Walker, 1971). Further, the liver of fish is found slow at hydrolysing even the BEE of 2,4-D, which is likely to accumulate there in detectable amounts, at least for some time. A fishing interval of three weeks can, however, ensure its disappearance from the fish liver. Safe fishing intervals must be determined for all herbicides and their formulations proposed to be used in the fishery waters. For dichlobenil this interval has been found to be as long as 60 days, while for endothall it is only three days.

Sometimes, **additives** contained in the trade products of herbicides may prove more toxic than the herbicide itself. This has been found true for diquat formulations containing a cationic additive. The absorption of tagged diquat by body of the fish was found very slow and inadequate to cause any fish mortality, but the additive was found absorbed by the fish body in fatal amounts. Such investigations on other herbicide formulations should prove very useful in practising safe use of herbicides in fish ponds and lakes.

9.6.: VEGETATION MANAGEMENT IN SWAMPS AND MARSHES

Our interest in swamps and marshes is in their wildlife comprising several kinds of aquatic and semi-aquatic birds like ducks, geese, coots, and waterfowl for hunting (Chapter-2). Therefore, the wildlife managers tend to encourage such vegetation as would provide these birds food and shelter. Presently, such birds are known to consume all parts of several grasses, sedges, and rushes, including their seeds, young foliage, inflorescence, and even underground stems. The seeds are rich in carbohydrates and, therefore, provide instant energy to birds for flying. *Scirpus, Cladium, Echinochloa, Leersia, Juncus, Fimbristylis, Cyperus, Zizania, Panicum, Spartina*, and *Phragmites* spp. are known to produce annually 150-900 kg seeds/ha. The ducks in marshes consume also specific plant parts of *Polygonums, Potamogeton, Lemna, Wolffia, Ceratophyllum, Pistia, Sagittaria*, and *Chara* spp. All such plants are to be encouraged and the unwanted trees, bushes, and other plants need be thinned out as per need, using largely physical methods.

It is equally important for the wildlife managers to be vigilant about the occurrence of any *non-native* plant specie which must be destroyed with the fastest acting means available on hand, at the first sight.

SELECTED BIBLIOGRAPHY

1. Ahmed, S.A. *et al.* 1982. Water quality as affected by waterhyacinth decomposition after cutting or 2,4-D application. Weed Res. Japan 27 (1): 40.
2. Beasley, P.G. *et al.* 1965. The absorption and distribution of ^{14}C-labelled diquat in gold fish (*Carossius airratus* L.). Proc. South Weed Conf. 18: 581.
3. Beresky. A. and S. Sarig. 1965. Use of aqua ammonia in fish ponds. Agri. Ammonia News, May-June, 55-57.
4. Blackburn, R.D. 1968. Weed control in fish ponds in US, FAO Fish Rept 44 (5): 1-17.
5. Canfield, D.E. *et al.* 1983. Effects of hydrilla and grasscarp on water quality in Florida lake. Water Resources Bull. 19 (5): 773-778.
6. Chokder, A.H. and A. Begum. 1965. Control of aquatic vegetation in fisheries. Agro. Pak. 16: 235-247.
7. Cope, O.B. *et al.* 1969. Effects of dichlobenil in the fish pond environments. Weed Sci. 17: 158-165.
8. Crause, J.H. 1963. The effects of copper sulphate on *Microcystis* and zooplanktons in ponds. Progressive Fish Culturist 25 (40): 198-202.

9. Engel, S. 1984. Evaluating stationary blankets and removable screens for macrophyte control in lakes. J. Aquat. Plant Manage. 22: 43-48.

10. Frank P.A. *et al.* 1967. Residues of two herbicides in water in irrigation canals. Weeds 15 (4): 353-355.

11. Frank, P.A. and R.D. Comes. 1967. Herbicidal residues in pond water and hydrosoil. Weeds 15: 210-213.

12. George, M.G. 1962. Occurrence of a permanent algal bloom in a fish tank at Delhi with special reference to factors responsible for its production. Proc. Ind. Acad. Sci. 56 B: 354-362.

13. Hiltibran, R.C. 1962. Duration of toxicity of endothall in water. Weeds 10 (11): 17-19.

14. Hiltibran, R.C. 1967. Effects of some herbicides on fertilised fish eggs and fry. Transactions Amer. Fish. Soc. 96 (4): 414-416.

15. Jacorossi, F. 1965. The toxicological problem of controlling weeds in canals. Weed Abst. 14, 1506.

16. Jhingran, V.G. 1968. Progress of inland fisheries research in India. Everyman's Science 3 (1): 38-47.

17. Kessler, S. 1960. Eradication of blue-green algae with copper sulphate. Bamidgeh 12 (1): 17-19.

18. King, S. 1996. Managing the blooming algae. IPM Practitioner 18 (7): 1-11.

19. Lawrence, J.M. 1957. Recent investigations on the use of sodium arsenite as an algaecide and its effects on fish production in ponds. S-E Assoc. Game and Fish Conm. 11: 261-266.

20. Link, K.H. and S.J. Low. 1991. Control of filamentous algae in an ornamental fish farm. Singapore J. of Primary Industries 19 (1): 40-49.

21. Martin, A.C. *et al.* 1957. Improving duck marshes by weed control. USDA Fish and Wild Life Service Circular No. 19, 60 p.

22. Mc. Lachlam, J. and P.R. Gorham. 1962. Effect of pH and N source on the growth of *Microcystis aerugimosa.* Canad. J. Microbiol. 8: 1-11.

23. Mitra, E. and A.C. Banrejee. 1968. Release of copper ions from pond soils. Ind. Agric. 12: 192-193.

24. Mitra, G.N. 1956. Reclamation of swamps for fishery development in Orissa. Progressive Fish Development (India) 54-62 pp.

25. Mullison, W.R. 1970. Effects of herbicides on water and its inhabitants. Weed Sci. 18 (6): 738-750.

26. Muniyappa, T.V. *et al.* 1995. Efficacy of postemergent herbicides on the control of waterhyacinth and their effects on fish. World Weeds 2 (2): 117-121.

27. Patnaik, S. 1972. Observations on the control of *Pistia stratiotes* in fish ponds by gramoxone. Proc. Ind. Acad. Sci. (B) 75 (1): 15-22.

28. Patnaik, S. and Ramachandran. 1976. Control of blue-green algal blooms with simazine in fish ponds. (*IN*) Aquatic Weeds in S.E. Asia. Dr. W. Junk b.v. Pub., The Hague, 285-292 pp.

29. Philipose, M.T. 1968. Present trends in the control of weeds in fish culture waters in Asia and the Far East. F.A.O. Fishery Report 44 (5): 26-52.

30. Philipose, M.T. *et al.* 1970. Some observations on the weeds of cultivable fish waters in Orissa. J. Inland Fishery Soc. of India, 11: 61-84.

31. Phillipose, M.T. 1963. Making fish ponds weedfree. Ind. Livestock (2): 20-21 and 34-36.

32. Ramachandran, V. *et al.* 1972. Observations on the use of ammonia for the control of *Pistia stratiotes* Linn. Inland Fish. Soc. 124-130.

33. Ramaprabhu, T. and V. Ramachandran. 1984. Developments in aquatic weed control research in India relating to fisheries. J. Aquat. Plant Manage. 22: 97-100.

34. Shafron, M. 1995. Blooming algae. Agr. Sci. 8 (1): 44-47.

35. Taylor, J.L. 1966. Casoron-a new aquatic herbicide. Hyacinth Contr. J. 5: 20-21.

36. Van Zon, J.C.J. 1984. Economic weed control with grasscarp. Tropical Pest Manage. 30 (2): 179-185.

37. Walker, C.R. 1963. Endothall derivatives as aquatic herbicides in fishery habitats. Weeds 11: 226 – 232.

38. Walker, C.R. 1964 (a) Simazine and other *s*-triazine compounds as aquatic herbicides in fish habitats. Weeds 12 (2): 134-139.

39. Walker, C.R. 1964. (b) Toxicological effects of herbicides on the fish environment. Water and Sewage Works Part 1 and 2, III (3), 8 pp.

40. Walker, C.R. 1964. Dichobenil as a herbicide in fish habitats. Weeds 22, 267-269.

41. Walker, C.R. 1965. Diuron, fenuron, monuron, neburon, and TCA mixtures as aquatic herbicides in fish habitats. Weeds 13, 297-301.

42. Walker, C.R. 1971. Problems in clearance and registration of herbicides for aquatic areas. Hyacinth Contr. J. 10: 5-17.

43. Welny, lP. 1968. Fertilisation of warm water fish ponds in Europe. F.A.O. Fish. Report 44 (3): 64-81.

44. Yeo, R.R. 1970. Dissipation of endothall and effect on aquatic weeds and fish. Weed Sci. 18 (2): 282-284.

Chapter 10

AQUATIC PLANT MANAGEMENT IN IRRIGATION SYSTEMS

Much of agriculture in India, and in many other countries of the world, would dwindle without a supply of water to crops to supplement rainfall, which is usually very erratic in nature. In India, since 1975 there has been an increasing trend to dam the perennial rivers and provide irrigation water to agricultural lands through network of canals. In good many instances the irrigation systems are accompanied by drainage ways to carry away the excess water, as part of the Command Area Development Projects. A variety of aquatic weeds is found to interfere with the easy flow of irrigation waters in canals and other water courses, causing reduced availability of water at distal ends and enhanced seepage losses. In drainage ditches the weeds cause complete choking and force waterlogging all around. Also, several filamentous algae clog the head-gates, screens, and irrigation siphon tubes (Chapter-1). These aquatic weeds and algae must be controlled in order to use our water courses efficiently. The agricultural-based economy in the country shall be noticeably affected by the spread of aquatic weeds.

10.1: DESIGN CONTROL DEVICES

Generally, in 3 m or deeper waters the growth of submersed weeds, as well as of certain rooted floating and emersed weeds, is kept in control because the sun light at that depth is not adequate to support plant growth. *Hydrilla verticillata* is one exception since it can grow in up to 4 m deep waterbodies. It is advisable to investigate into more of such interactions of individual aquatic plant specie and variable water depths and, when possible, design the waterbodies narrow and deep. Similarly, in canals and other water streams, the fast flowing waters discourage the under-water weeds by making the water turbid which hinders penetration of light through it. The slow streams, on the other hand, allow the silt to settle and thus keep the water clear. Both these features, *i.e.* the silting of the waterbody and clean water, encourage heavy weed growths. While it is true that it is not always possible to design waterbodies deep and fast flowing solely to keep the aquatic weeds away, yet a compromise to aim at overall best performance of the waterbody is always possible.

Lining of canals, both at the bottoms and on sides, has been followed in many canals, again primarily for structural reasons although it is supposed to give added value benefit of suppressing aquatic weeds and their allies. But, unfortunately, none of the linings, plastic, concrete, or tile, has yielded desired weed suppression effects. While the plastic linings are easily penetrated by weeds, other linings permit aquatic weeds through joints and cracks. Besides this, gradual deposition of silt in the lined canals support luxuriant growths of aquatic weeds. Then there are weeds like *Phragmites* spp. which start in shore waters, but then spread to the lined banks with the help of their long, creeping rhizomes. The lining eventually becomes weak and ineffective in checking further weed growths.

10.2: PHYSICAL CONTROL DEVICES

Physical methods are used almost exclusively in many countries to weed out canals, their branches, distributaries (both major and minor), and waterways. In India, large labour force is deployed each summer during the hottest part of the year for digging the canal bottoms, primarily to remove their silt deposits. Although at this time the demand of agricultural water is supposed to be at its minimum (except in cotton and sugarcane crop areas), yet the drinking water supply is often put under great stress. The canal closure operation continues for about 5-8 weeks.

Desilting of canals is essential to revive their flow capacities each year. During the process, numerous aquatic weeds are also removed from the canal system. Also, the dug-bottoms expose the shallow weed propagules for desiccation under the hot sun. But all this proves insufficient to suppress aquatic weeds since within 2-3 weeks after the release of water in the canals, the weeds show-back and eventually grow to their original levels.

It may be further observed here that slowing down of water-flow by weeds in canals and other streams aids in settling of the silt otherwise suspended in the fast flowing waters. The silty water bottoms, in turn, encourage additional growth of rooted aquatic weeds, and this vicious circle continues to our disadvantage. Thus, round the year efforts of cutting and raking of aquatic weeds, particularly of the submersed ones, become the primary requirement in the management of our streams.

Mechanical methods offer the best choice for weed control in particularly large canals of over 2.8 cumsec discharge (see Chapter-6).

10.3: BIOLOGICAL CONTROL OPTIONS

Grasscarp is an extremely useful bioagent for the control of submersed aquatic weeds, especially in the high discharge canals and other large waterbodies. Alongwith mechanical methods it has proved a boon for the clearance of very large canals in some countries. In such high discharge canals herbicides cannot be employed against submersed weeds. A very good study on the comparison of acrolein and grasscarp in a multipurpose river in Egypt is reported by Van Zon (1984).

Floating weeds like waterhyacinth and salvinia are now controlled efficiently with certain arthropod bioagents already described in Chapter-8.

10.4: HERBICIDAL CONTROL OPTIONS

Herbicidal control is usually considered for comparatively smaller canals since the quantity of herbicide needed in large canals is often too high to be economical. For instance, to treat a canal of 5.3 cumsec with a herbicide at 0.5 ppm, it is required to add 120 kg *a.i.* of the herbicide per km length of the canal. But in the low discharge irrigation canals and in limited sized waterbodies certain herbicides can hold back the weeds economically for a season or longer.

Not all herbicides can be used in irrigation waters. Certain important features of a suitable herbicide for treating irrigation systems are:-

1. It should show sufficient phytotoxicity to aquatic weeds so as to kill them in acceptable contact time.

2. It should dissipate from the treated water soon after damaging the weeds so that the water is left safe for irrigation.

3. It should be easy to apply to the flowing waters.

4. It should possibly be safe to fish and other useful aquatic fauna since in the developing countries canals are always double purpose waterbodies, supplying irrigation water as well as fish and crustaceans.

Unfortunately, no aquatic herbicide with all the above four features is yet developed and one has to choose the best from the available compounds for a given ecosystem.

10.4.1: Common Herbicide Options

(i) **Emulsified solvents:** Emulsified solvents are aromatic oils containing suitable emulsifiers. These are very potent herbicides to control submersed aquatic weeds in flowing waters ***where fisheries is not important***. These inflict a quick, contact injury on the weeds and then leave the water rapidly. These do not cause any important injury to crops when the treated water is used for irrigation (Frank, 1968; Bruns and Yeo, 1964). A common application rate of emulsified solvents is 1623 l per cumsec flow of water or 45 l for every 30 cm width of canal, whichever is more. The injection time* of emulsified solvents is half to one hour. Treatment from the main station is effective up to 4 or 5 km length of the canal after which a booster injection with half the dose of the solvent is required. The emulsified solvents prove most effective when these are applied to the weedy canals at the first report of their interference with the water flow. After the submersed vegetation has fully grown and reached the water surface, it shall create difficulty in an even distribution of the emulsified solvents through the length of the canal. The same holds good for the already clogged drainage and other ditches.

Emulsified solvents are usually good for treating the fast moving or churning waters. In the slow moving or still waters their aromatic contents tend to surface out. The optimum flow rate of water for treatment with emulsified solvents in canals is 0.17 to 0.42 m sec^{-1}. Silty waters are unsuitable for treatment with emulsified solvents. Also, in very cold waters the emulsified solvents oil-out while in the very warm waters they are subjected to rapid volatile losses. Water temperature of 15 to 32°C is considered suitable for the application of emulsified solvents to the flowing waters. Under unfavourable water temperature conditions one way to use the emulsified solvents effectively is to increase their emulsifier content to 2 to 3%, instead of the usual 1 to 1.5%.

Emulsified solvents are considered of low toxicity to humans. ***But these may irritate eyes and skin. Fish are killed at 10-15 ppm. Their fumes may be toxic to waterfowl. The treated water should not be used for cooking food.***

(ii) **Acrolein:** It is more potent a herbicide than emulsified solvents against submersed and floating weeds and algae. It is used in canals, ditches, ponds, and other **non-fishery** irrigation water systems.

** *Note: Injection time is the duration of continued injection of a herbicide in a flowing waterbody.*

Acrolein is soluble in water, therefore, it does not require assistance of any emulsifier. For the same reason it is more suitable than the emulsified solvents for treating long and slow moving channels where the latter would tend to oil-out. A single treatment with acrolein at 162 to 486 l per cumsec of water flow can control the submersed weeds for a distance of up to 8 km in small channels and up to 32 km in large canals. One or two applications of acrolein each season will prevent excessive growth of aquatic plants from accumulating in canals. Lower application rates are recommended in warm and slow moving waters, while higher rates are used in the cool and fast moving waters.

The common injection time of acrolein ranges from 1 to 4 hours without vitiating the results.

In India acrolein was successfully used in a sub-branch of Chambal canal in Rajasthan (discharge 0.056 cumsec). Acrolein was injected for 4-5 hours at 50 l/hr. It killed weeds up to 5 km down the stream for 12-14 weeks. But in portions of the ditch infested with dense growth of pondweeds (*Potamogeton* spp.), acrolein movement was hindered and so it proved ineffective (Mehta, 1975). *

Acrolein is injected by special equipment into water, about 30 cm below the surface to form a wave of treated water. ***It is very toxic to fish. It irritates eyes, skin, and the respiratory system. It also causes chemical burns. Therefore, acrolein is used under the supervision of trained and licensed operators.***

Cotton is extremely susceptible to the acrolein-treated irrigation water.

Do not permit dairy animals to drink the acrolien-treated water for 2-3 days after treatment.

(iii) Copper sulphate (CSP): Copper sulphate is used in irrigation waterbodies to kill several species of algae, including chara. It is usually applied at 0.1-11.0 ppm, but sometimes its repeat application at low doses for 3-5 days proves more effective than its one time, full dose application. In flowing waters its rate is 5.30 kg/cumsec discharge for scum algae control and up to 64 kg/cumsec for the control of *Chara* spp. This quantity of CSP is sufficient up to 6-10 km length of channels whereafter a booster application is usually needed. Rhodamine B dye can be applied alongwith CSP to trace its arrival at the booster point. *Excessively high rates of CSP may be* *toxic to fish.* CSP should be applied as soon as algae are observed otherwise they will have opportunity to produce millions of spores for their further survival.

In irrigation waters CSP is, at the present, most economical algicide. The treated algae turn greyish-white soon after treatment. The water will be free of algae within seven days. It must be applied on sunny days to obtain effective results. Algae control should last 5 weeks or longer. Swimming and fishing are allowed immediately after the treatment. Do not apply CSP to water at below 15°C temperature.

(iv) **Dichlobenil:** It is a short persistence herbicide, used largely for the exposed canal bottom treatment. A number of aquatic weeds, including pondweeds and naiads, are susceptible to it. The common rates for hydrosoil incorporation are 5-10 kg ha^{-1}. The treated hydrosoil is left exposed for about five weeks before the water is let in. The canals can then be used 24 hours later for supplying irrigation water to crop fields. During this interval the dichlobenil residues are diluted to negligible levels. Soil incorporation of dichlobenil can keep the channels free of submersed weeds for a season or more.

(v) ***Diquat:*** Diquat can be used in all kinds of irrigation waterbodies, including farm channels, to control floating and emergent weeds. However, the shoots of rooted perennial weeds will regrow after sometime. Its common rates of application are 1-2 kg/ha.

Diquat can also be used against submersed weeds and algae. It can be combined with CSP to increase the phytotoxicity of copper to hydrilla (Chapter 7). *The diquat-treated water can be used for irrigating crops about 10 days after the treatment.*

(vi) ***Diuron:*** Diuron can be used for treating the soil of irrigation ditches when water can be made to stand in it for 72 hours after the treatment. Drain off and waste the left-over water before using the ditch for supplying irrigation water.

(vii) ***Endothall:*** Granules of endothall are used in confined irrigation waterbodies against many submersed macrophytes and lemnids. The herbicide has a short persistence of 7-15 days. The effective rates are 0.5 to 5.0 ppm (Chapter 7). Water should be at 16°C or more for obtaining best results.

(viii) ***Anhydrous ammonia:*** Anhydrous ammonia leaves the treated water safe for irrigation after disintegrating the submersed, noxious vegetation. *Hydrilla* and *Najas* spp. are partially susceptible to it. The effective dose of ammonia is 10 ppm or more. ***It is very toxic to fish.***

(ix) ***Glyphosate:*** At 1-3 kg/ha (or as 0.1-0.3% aqueous spray), glyphosate is permitted for control of weeds in drainage ditches, ditchbanks, and like situations. Also, inside waterbodies it has been found effective in localised treatment against waterhyacinth and water-lily.

(x) ***Metham (SMDC):*** Metham forms a strong fumigant. In Irrigation systems it is recommended for freeing the large, underground **drainage pipes** of plant roots entering there from the ground and choking their water-flow. In such pipes, metham can be applied either by soaking it as liquid through the cracks and openings visible on the ground or by pumping-in its air-aqueous foam formulation. In the case of tree roots entering the pipes, the concentration of metham should be so chosen as would kill only the roots entering the pipes and not the tree itself. A common concentration used for the purpose is 1450 ppm. But when it is desired to kill the roots of total vegetation inside the pipes, of course, higher concentrations of SMDC can be used.

SMDC eventually is lost by vapourisation, leaving the drainage water unpolluted for release at the other end.

Methods can be devised also to treat the above ground pipes choked with vegetation with metham. Metham is not hazardous to human life.

10.5: EFFECT OF HERBICIDE RESIDUES IN IRRIGATION WATER ON CROPS

The aquatic weed scientists have reported some useful field experiments aimed to study the effects of herbicide residues present in irrigation waters on the growth of crops. In general, these studies revealed that the herbicides recommended for use in irrigation water were safe to most of the crops irrigated with the treated-water when their prescribed methods and rates of application were adhered to. Also, higher herbicide residue concentrations were tolerated when the crops were irrigated by furrow method than by the sprinkler method .

10.6: MANAGEMENT OF DITCHBANK VEGETATION

Although dense vegetation growing on the ditchbanks is unwanted, a limited growth of the low growing grasses like *Cynodon dactylon, Eragrostis, Poa,* and *Paspalum* spp., is always desirable since it helps in preventing bank erosions without causing any kind of obstruction to the flow of water. Their fibrous roots bind the soil, in variance with the tap-rooted plants which often tend to create cracks in the ditchbanks. Tall grasses, despite their fibrous roots, are of course undesirable since they badly hinder approaches to the waterbodies, besides hiding rats and snakes (Chapter 2).

Under Indian conditions, **grazing** by different kinds of animals, including cattle, sheep, goat, and horses, is a very economical way of keeping the ditchbank vegetation trimmed to desired levels. But sometimes, this biological approach to manage the ditchbank vegetation can give rise to selective growth of unpalatable, thorny, and woody weed species, which may be more cumbersome than the original ditchbank vegetation. In some places *Parthenium hysterophorus* is reported to have spread on ditchbanks just because most animals refused to graze upon it. But the author witnessed beautiful banks of Gang Canal in Rajasthan covered with surface growing vegetation, holding the soil *in situ* and providing an appealing look. Animal grazing alone achieved this year after year. The grazing has been non-selective, at least in this area. However, every where it can not be so and one may have to integrate animal grazing with occasional **mowing** with either manual sickles and scythes or mechanical mowers. The latter work only on evenly sloped ditchbanks besides the fact that the stout weedy plants require very high H.P. mowers to cut them. Also, even close mowings can still leave some thorny or spiny surface creepers to spread as troublesome plants.

Certain **herbicides** are successfully employed to supplement other options of ditchbank weed management, where needed. Herbicides are particularly preferred as initial aids for clearing the neglected banks of their old, stout, vegetation. Also, weeds that escaped grazing and mowing of the ditchbanks, can be spot-treated with specific herbicides. Dalapon, TCA, MSMA, asulam, amitrole, paraquat, glyphosate, MH, 2,4-D, fenoprop, buthidiazole, and herbicidal oils form common ditchbank herbicides. The choice amongst these can be made keeping in view the target weed species (Table 7.2). The killed vegetation should then be put to fire to allow the low growing grasses to take over.

Often doubts have been cast on the **non-target effects** of herbicides applied to ditchbanks. Sometimes these may reach the main body of irrigation water through spray drifts and/or runoff from the treated banks. Spilling of herbicides may be another source of contamination of water, which of course is more a part of carelessness than the default of the herbicide application.

Of the several recommended ditchbank herbicides, 2,4-D has been implicated most in damaging at least some crops, particularly the fruits and vegetables. Adequate care should be taken to minimise its spray and volatile drifts reaching the waterbody by accounting for the wind speed, wind direction, atmospheric temperature, and the kind of spray nozzles used for its application. Herbicide run-off from the treated banks can be avoided by using low volume sprays and avoiding their application during the rainy period. Finally, the recommended waiting periods should be strictly adopted before using water from the bank-treated waterbody for irrigating crops.

In the flowing waterbodies most of the ditchbank herbicides that reach the streams are diluted rapidly after the water has travelled some distance. The major herbicide contamination problems warranting precautions lie in treating the banks of confined irrigation waterbodies, like the lakes and ponds.

10.7: MANAGEMENT OF WEEDS IN DRAINAGE DITCHES

In the new irrigation projects, drainage ditches are provided to protect the surrounding area from waterlogging and save numerous hectares of cultivated land from turning into waste lands. But several tall weeds, particularly *Typha* spp., grow luxuriantly in these drainage ditches and block them completely (Chapter-1). Other common weeds acting the same way in different parts of the country include *Phragmites*, *Sorghum*, and *Ipomoea* spp. Physical **cutting** of such weeds has often met with frustration because of their rapid regrowth from networks of rhizomes. In recent years **herbicides** like paraquat, with or without 2,4-D; dalapon with or without paraquat; TCA, MSMA, buthidiazole, asulam, amitrole-T, herbicidal oils, and glyphosate have been found effective against such weeds in the drainage ditches. The dead weeds can then be easily pulled out from the ditches to let the drainage water pass through. There is a view that the recovered ditches can be usefully transplanted with some low growing grasses which will not hinder the water-flow, and at the same time discourage the regrowth of *Typha* and other such tall weeds. Some scientists further believe that during construction of the new drainage ditches, the soil could be sterilised with soil sterilants like diuron, TCA, simazine, bromacil, fenac, sodium chlorate, or fenuron. But their use is often limited by the fact that water from the treated drainage ditches may cause environmental hazards on the way which is sure to invite problems.

SELECTED BIBLIOGRAPHY

1. Balyan, R.S. and D.S. Malik. 1982. Pearlmillet and its sensitivity to aquatic herbicides. Ind. J. Weed Sci. 14 (1): 31-33.
2. Barrett, P.R.F. 1981. Diquat and sodium alginate for weed control in rivers. J. Aquat. Plant Manage. 19: 51-52.
3. Blackburn, R.D. and L.W. Weldon. 1964. Control of Southern Naiad in Florida drainage and irrigation channels: Weeds 12: 295-298.
4. Bowmer, K.H. and G.H. Smith. 1984. Herbicides for injection into flowing water: acrolein and endothall amine. Weed Res. 24 (3): 201-211.
5. Bowmer, K.H. and G.R. Sainty. 1977. Management of aquatic plants with acrolein. J. Aquat. Plant. Manage. 15: 40-46.

6. Brenzy, O. and I. Mehta. 1970. Aquatic weeds and their distribution in Chambal Command area in Kota (Rajasthan). Ind. J. Weed Sci. 2 (1): 70-80.
7. Brenzy, O. *et al.* 1976. Control of weeds in irrigation channels. Ind. J. Agro. 21 (2): 173-176.
8. Bruns V.F. and R.R. Yeo. 1964. Tolerance of certain crops to several aquatic herbicides in irrigation water. ARS, USDA Tech. Bull. No. 1299, 1-22 pp.
9. Bruns, V.F. and A.D. Kelley. 1974. Effect of sprinkler irrigation with xylene-treated water in six crops. CARC, Washington State Univ. Bull. No. 796, 4 p.
10. Bruns, V.F. *et al.* 1972. Response of several crops to six herbicides in irrigation water. Tech. Bull. 1461, Agr. Res. Sta., USDA, 29 pp.
11. Burdrick, G.E. *et al.* 1964. Toxicity of aqualin to fingerling brown trout and blue gills. N.Y. Fish Game J. 11: 106-114.
12. Canfield, D.E. 1983. Sensitivity of Florida lakes to acidic precipitation. Water Resources Res. 19 (3): 833-639.
13. Comes, R.D. and A.D. Kelley. 1979. Response of certain crops to glyphosate in irrigation water. Weed Sci. 27 (6): 658-660.
14. Comes, R.D. *et al.* 1984. Response of concord grapes (*Vitis labrusea*) to 2,4-D in irrigation water. Weed Sci. 32: 455-59.
15. Couch, R.W. 1990. Managing algae growth with grasscarp and algicides. Weed Abst. 42 (6), 2587.
16. Crossland, N.O. and W.M. Adams. 1963. The use of acrolein for the control of aquatic weeds and snails in an irrigation system in Tanganyika. PANS (C): 9: 221-224.
17. Demint, R.J. 1970. Amitrol residues and rate of dissipation in irrigation water. Weed Sci. 18 (4): 859-864.
18. Dutta, T.R. *et al.* 1972. Evaluation of herbicides for submersed weeds in Chambal and Bhakra--Nangal Canal systems. Ind. J. Agri. Sci. 42 (1): 70-75.
19. Engel, S. 1984. Evaluating stationary blankets and removable screens for macrophyte control in lakes. J. Aquat. Plant Manage. 22: 43-48.
20. Frank, P.A. 1968. Analysis of emulsified xylene in water. Weed Sci. 16 (4): 489-490.
21. Frank, P.A. *et al.* 1963. Evaluation of herbicides applied to soil for the control of aquatic weeds in irrigation canals. Weeds 11: 124-128.
22. Frank, P.A. *et al.* 1970. Herbicides in irrigation water following canal bank treatment for weed control. Weed Sci. 18 (6): 687-692.
23. Haller, W.T. 1981. Maintenance control of waterhyacinth. Aquatics 3 (2): 6-7, 11-12.
24. Hodgson, J.M. 1968. Chemical control of seed canarygrass on irrigation canals. Weed Sci. 16 (4): 465-467.
25. Hodgson, J.M. *et al.* 1962. Control of certain ditchbank weeds in irrigation system. Agri. Res. Sta., USDA, Production Res. Rept. No. 60: 1-64.

26. Mehta, I. 1975. Some experiments with xylene, acrolein, and Hydrothol-191 for the control of submersed weeds. Andhra Agri. J. 22 (1 & 2): 20-24.

27. Mehta, I. and R.K. Sharma. 1978-1979. Control of aquatic weeds in Chambal irrigation system by weed cutting launch. Agr. Engineer 21 (3): 39-40.

28. Millhollow, R.W. 1969. Control of Johnsongrass in drainage ditchbanks in sugarcane. Weed Sci. 17 (3): 370-373.

29. Nelson, J.L. *et al.* 1969. Behaviour and reactions of copper sulphate in an irrigation canal. Pesticide Monitoring J. 3 (Dec.): 186-189.

30. Pearce, G.A. 1965. Weed control in irrigation channels. J. Agri. (West Australia) 6: 94-104.

31. Robinson, G.W. 1969. The use of herbicides in the maintenance of land drainage channels on Romney Marsh. J. Inst. Water Enggrs. 23: 159-176.

32. Sinclair, L.R. 1980. Nutrient removal from drainage waters with systems containing aquatic macrophytes. Trans. Amer. Soc. Agr. Engg. 1189-1194.

33. Singh, D.J.C. and K.N. Rao. 1976. Aquatic weed control in irrigation channels and drain of Andhra Pradesh (India). Andhra Agr. J. 26 (1 & 2): 1-9.

34. Toscani, H.A. 1983. Progress in the control of submersed weeds in dams. Malezas 11 (5): 232-260.

35. Van Zon, J.C.J. 1984. Economic weed control with grasscarp. Trop. Pest Manage., 30 (2): 179-185.

36. Vander L.L. *et al.* 1960. Weed control in dams and irrigation channels. Rohd. Agric. J. 57 (5): 350-358.

Elodea nuttallii (Waterweed)

Chapter 11

NOXIOUS AQUATIC PLANTS AND THEIR CONTROL

1.[*] *Eichhornia crassipes* (Mart.) Solms (waterhyacinth) **(Fig. 3.7)**

Biology and Habitat

Waterhyacinth, a native of South America, is the most gregariously growing aquatic weed in 52 tropical and subtropical countries of the world, including India. In India, it was introduced in Bengal in 1886 as an ornamental pond plant from Brazil. From Bengal it escaped and spread as an aquatic weed throughout N.E. India, part of which is now Bangladesh.

Today, waterhyacinth covers about 2 million ha of water area in India, chiefly in the states of Bihar, West Bengal, Assam, Uttar Pradesh, Andhra Pradesh, Tamil Nadu, Orissa, and Kerala. Fish and rice crops worth millions of rupees are lost every year at the hands of this menace. A rapid vegetative growth through its offsets, particularly during the monsoon season, is the main secret of its quick build-up into massive mats, strong enough to stop even large boats and ships. Two waterhyacinth plants have been recorded to grow into 1200 offsprings in 120 days. Center and Spencer (1981) estimated in Florida

---

Note:

1. *The weeds are arranged here by alphabet and not by their any relative importance which differs from place to place.*
2. *The choice of any herbicide against any weed included in this chapter should be based on guidelines already provided in Chapters 7, 9, and 10.*
3. *The plant descriptions and sketches are given in Chapter-3.*

biomass accumulation rate of waterhyacinth as about 20 g/sq m/day. The peak efficiency of tissue storage relative to solar input was approximately 1.4%. The optimum water temperature for its growth was observed to be 28-30°C. Hyacinth infestations of a meter height, tightly bunched structure, weighing as much as 500t/ha are not uncommon.

Besides a preponderate vegetative reproduction, copious seed production in waterhyacinth enables the weed to encroach upon new areas and recolonise the old ones. The waterhyacinth seeds remain dormant for several years in deep waters, but they sprout in shallow water situations where adequate light and O_2 were readily available. Thus, shallow ditches and shore water areas are preferred sites for the establishment of the hyacinth seedlings in a submersed state. These seedlings strike roots in the substrata and continue to put forth new leaves. The sixth leaf on each seedling is buoyant because of its abundant aerenchyma tissue. At this stage the hyacinth seedlings break off and rise to the surface of water. The floating seedlings then produce new fibrous, adventitious roots and continue to add more leaves. The plant is now independent and is ready to go into deep, main waters with the water currents. There they develop buoyant petioles which are usually short and swollen (2-3 cm in diameter), but sometimes these may grow long, exceeding 100 cm in length. The inclination of hyacinth to grow long petioles is generally noted in polluted waters and overcrowded situations of confined waterbodies.

Waterhyacinth is capable of striking roots in soil during the period the ditches go dry.

Man and his animals are the main agents of dispersal of waterhyacinth. The hyacinth-infested, shallow waterbodies are often dug for using their bottom soil as manure for agricultural fields. In rivers and canals, hyacinth mats are dispersed with boats, wind, and water currents. Livestock pick up waterhyacinth seeds in their hooves while moving about in the infested waterbodies.

Waterhyacinth compost often spreads its inconspicuous seeds and live rhizomes through farm channels and paddies. Too, the waterfowl and other shore birds have been found to carry hyacinth seeds during their migration to long distances.

Control Options

Manual lifting of hyacinth mats is a common practice in India.(Plate 11.1). Early rainy season, before the hyacinth mats add new dimensions, is an appropriate time for this operation. Depending upon the density of the weed mat, it may require 100-300 man days to remove the weed from one ha of waterbody (Philipose 1963; Ramachandran, 1969). In West Bengal and some Eastern and Southern states of India, special hyacinth destruction weeks are often organised. In Uttar Pradesh, manual removal of waterhyacinth was encouraged by providing monetary assistance for turning it into compost. In the coastal areas of the country, dredging of waterhyacinth mats with boat-mounted nets into the sea is practiced, where it dies rapidly.

The entry of hyacinth mats into new waterbodies is often checked physically with strong bamboo screens, although when such screens give way, overnight disasters in terms of massive movement of waterhyacinth may occur. Heavy infestation of waterhyacinth in Keetham lake in Agra (India) was the result of the historic tumbling down of the Okhla hyacinth barrier.

In western countries mechanical hyacinth destroyers have been developed and used on large scale to collect, convey, and even crush the hyacinth mats in sequence. In India certain indigenous, portable machines for removing waterhyacinth and like free-floating plants were developed but these did not find wide field acceptance.

Several **uses** of waterhyacinth have been proposed but so far only two of these look promising. These are: (i) composting and (ii) biogas production (Chapter 6). Besides, there is potential of turning waterhyacinth into palatable silage. It can be done by dewatering the plants to 50-90% moisture and adding 4% dried citrus or cracked maize and 1% cane molasses before filling the silage pits. Chibbar and Singh (1971) reported that even fresh waterhyacinth plants could be used for making silage by mixing 25% rice straw. As an animal feed, waterhyacinth can at best replace 25% of daily animal ration. It is so because the plant is not so platable. Besides, it is high in mineral content and may also possess some undesirable constituents (Chapter-6).

Herbicidal control of waterhyacinth has made considerable advancements. Numerous researches have established the usefulness of **2,4-D** (and MCPA) in killing and sinking the waterhyacinth mats *in non-agricultural waters.* The esters of 2,4-D have provided better weed kill than its sodium or amine salts, but the former are more prone to drift hazards, particularly during their aerial application. 2,4-D is now the world-wide used herbicide against the hyacinth menace (Plate 11.2). Its application rates vary widely from 1 to 8 kg ha^{-1}, depending upon the density of the mat and several other factors, including the feasibility factor. However, a dose of 2 kg 2,4-D ha^{-1} is most frequently used. The best time to treat the hyacinth mats with 2,4-D (or MCPA) is during warm weather when the weed has put forth some new growth. The treatment may have to be repeated 2-3 times by spot application to kill the escapes. By aerial application, ester 2,4-D is found effective at as low a dose as 0.36 kg ha^{-1}. A strong sticker should be added to 2,4-D to improve its herbicidal effects and also to minimise its volatile and drift hazards. The 2,4-D treated hyacinth mat is browned within a weak and it sinks into water in another 6-8 weeks.

Plate 11.1 : A large infestation of waterhyacinth being helplessly tackled by men somewhere in Andhra Pradesh. (Courtesy – DJCS and KNR)

Plate 11.2: A confined waterbody filled with waterhyacinth being treated with a herbicide using a hand sprayer.

Plate 11.3: Close-up of an infestation of waterhyacinth severely attacked by its bioagent, *Neochetina eichhorniae* Warner (Photo-Courtesy NRC (WS), Jabalpur)

Plate 11.2

Plate 11.3

Pieterse and Roorda (1982) reported that if the hyacinth mats were preconditioned with a growth retardant like chlorflurecol (45 g/ha) or gibberellic acid (6 g/ha), 2,4-D could damage these at a much lower dose. But some authors did not find it an economically viable proposition (Joyce and Haller, 1984) although the smaller dose of 2,4-D may reduce the ill effects of the herbicide in water on the non-target organisms. In situations where 2,4-D is not permitted to be added in waterbodies even in minute quantities, like in agricultural waters, **diquat** (1-3 kg/ha) has been employed as, a contact, quick acting herbicide against waterhyacinth. In drainage ditches and marshes, **amitrol-T** (1-2 kg/ha) has effectively destroyed the hyacinth growths. In some other waterbodies the hyacinth mats have been successfully sunk with **ametryn** (at 3.5 kg/ha) without any residual effect in the treated waters. Where permissible, **a mixture of 2,4-D and paraquat** can kill this weed more efficiently than either herbicide applied alone. In Argentina, **hexazinone** (at 2.25 kg/ha) has been reported to give a long lasting control of waterhyacinth for up to 12 months against the already recommended herbicides which keep back the weed for only up to 2 months.

Biological control of waterhyacinth is now a reality and, in some cases, it offers more promise than herbicides which have become so expensive, besides being very situation specific. The technique was first given serious attention in 1960 but its field application got momentum in 1968 at the hands of Dr. B.D. Perkins and Dr. C.J. De Loach of USDA. At the present, three insect species, all native of Argentina, have been successfully released in the USA and elsewhere against this weed.

*These insects are:

1. *Neochetina eichhorniae* Warner, (Curculionidae)

2. *N. bruchi* Hustache, (Curculionidae)

3. *Sameodes albiguttalis* (Warren). (Pyralidae)

The first two insects are semi-aquatic hyacinth weevils, while the third one is a moth. In Florida (USA) These insect bioagents are now well established on waterhyacinth for over a decade.

The young larvae of the two hyacinth weevils burrow under the leaf epidermis and work their way towards the base of the leaf, or sometimes to the base of young petioles. (Plate 11.3). When fully grown, these larvae burrow through the stem and move out to the upper

---

* *CHLORFLURECOL = 2-chloro-9-hydroxyfluorene-9-carboxylic acid methyl ester.*

root zone just under the water surface. The extensive damage of vascular tissues thus caused may lead to wilting of the affected leaves and, in extreme cases, the petioles may be severed from the stems. Slowly the hyacinth carpet is reduced in height, followed by inhibition of its growth (Plates 11.4 and 11.5). But, the hyacinth weevils being slow at their reproduction, the survivorship of the dense waterhyacinth mats is little affected. Van and Center (1994) found that the efficacy of hyacinth weevils in controlling waterhyacinth could be improved if the hyacinth mat was pre-treated with a growth retardant like paclobutrazol at 2.2 kg/ha. Also, Visalakshy and Jaynath (1996) observed that in shallow waterbodies and on shores where the hyacinth roots may hit the ground, the efficacy of the two weevils is greatly reduced.

Where the waterbodies and their hyacinth infestations are of seasonal nature, it is imperative to release sufficiently large number of *N. eichhorniae* and *N. bruchi* to kill the weed within its growing season before the waterbody dries up. This approach of biocontrol is called `inudative approach'. Kathiresan (1999) reported success of this approach against waterhyacinth by releasing 3 insects per plant on the small hyacinth plants and double the population on the medium-sized plants. In case of large plants, their leaves being unpalatable to the insects, this bioagent cannot act quickly at any release rate.

Sometimes heavy downpour of monsoon can fail an aquatic biocontrol operation by washing away the insect bioagent.

Contrary to the two hyacinth weevils, the hyacinth moth (*S. albigutallis*) builds-up its population fast enough to curb the survivorship of waterhyacinth (Center, 1982, 1984; Centre and Durden, 1981). Its larvae, which are about 2 cm long, with a dark orange head and a cream colour body, feed severely on very young, central leaves and apical buds of waterhyacinth. The centres of the affected rosettes are wilted or browned, The actively growing hyacinth mats show the greatest damage because of the presence of many new, young leaves and buds on them. After inflicting considerable damage to the young leaves, the larvae seek out fairly large sized petioles and burrow into them, making tunnels for pupation. These larvae usually prefer inflated, float type petioles for this purpose. The damage is seen as small, dark brown spots on the hyacinth petioles. All this leads to fast disposal of the weed. The moth was first released in Florida (USA) in 1977 where the scientists observed the damage rate of about 4 km long hyacinth carpet per day (Center and Durden, 1981; Center, 1984). It is important that only disease free colonies of bioagents are released on the hyacinth mats.

Plate 11.4: A waterhyacinth-infested pond.

Plate 11.5: The pond in Plate 11.4 about two years after release of *Neochetina eichhorniae* as a bioagent of waterhyacinth. Note the weedfree water in the rear. (Photo-Courtesy NRC (WS), Jabalpur)

Thus far, the hyacinth moth (*Sameodes albiguttalis*) has not been allowed to be imported for testing in India for unknown reasons.

In India, the two hyacinth weevils (*Neochetina* spp.) have been released for field testing in different states. In many places the programme had been greatly successful. A report from Jorhat (Assam), for instance, showed that in the Disangmukh area of Sisbar district over 1000 ha of waterbodies are cleared of waterhyacinth by field release of *N. eichhorniae* and *N. bruchi*. Now these areas are used for paddy cultivation. Similarly in Jaisagar area of Sibsagar district, waterhyacinth has been brought under control the biological way. The two weevils have dispersed through rivers as well as through aerial migration to the six districts of Assam, *viz.*, Sibsagar, Dibrugarh, Golaghat, Nowgong, Kamrup, and Sonitpur. Here the growth of waterhyacinth has started stunting and the flowering has become less dense. Similar reports are available from Jabalpur (M.P.), Banglore (Karnataka), and some other places.

2. *Hydrilla verticillata* (L. F.) Royle (hydrilla) **(Fig 3.23)**

Biology and Habitat

Hydrilla is the most dominant rooted, submersed weed in India, infesting ponds, lakes, canals, and, sometimes, also the rivers. In variance with other submersed weeds it is capable of infesting even up to 12 m deep waters. Obviously, its light and CO_2 compensation points are very low. Bowes *et al.* (1977) showed that hydrilla plants were adapted to as low as 10-12 einstein/sq m/sec light energy and 50 ml CO_2/l water surrounding it.

Hydrilla shoots ramify fast into dense masses to completely fill the lakes and ponds. Its growth is particularly luxuriant in calcareous hydrosoils and in high pH waters. Many fishery waterbodies have been abandoned in the country because of invasion by hydrilla. The decaying masses of worn-out hydrilla rot and produce highly offensive odours. In canals, hydrilla heavily cuts the water flow rates.

The presence of turions in hydrilla, particularly the subterranian ones (called tubers), makes this weed highly persistent (Chapter-3).

Control Options

Physical methods can be employed to mechanically mow hydrilla, but the mowers do not reach the plants below 1 m water depth. The detatched, floating masses of hydrilla shoots can then be collected manually down the stream near some bridge arches. In the western world, the collection of hydrilla has also been mechanised. But in every case physical mowing and collection of hydrilla are periodic efforts, not only to keep the weed in check but also to prevent dissemination of its cut shoots.

Certain **herbicides** can substitute mowing of hydrilla without fear of causing its dissemination. In Orissa (India) **anhydrous ammonia** was found to effectively disintegrate hydrilla shoot masses. For this, ammonia is released about 60 cm deep in water to create a nitrogen concentration of 18 ppm. The hydrilla nets begin to give way within five days of the treatment. The anhydrous ammonia treatment is suited to the renovation of fish ponds (Chapter 9) and treating other waterbodies *not meant for fish cultivation* (Ramachandran, 1960 and 1969).

In fishery waters a **combination of diquat and CSP** (or cutrine) has given excellent destruction of hydrilla. Probably, diquat induces extra accumulation of copper in hydrilla tissues which aggravates phytotoxicity to the weed. The recommended concentrations of the two components are 1 ppmv diquat and 1-4 ppmv copper ion. For achieving this usually 3-4 kg/ha of diquat and 1-3 kg/ha of copper compound are required. In muddy waters, invert emulsion of diquat and Cu complex should be preferred to avoid their inactivation due to the adsorption of diquat on particulate matter. The diquat and Cu complex must be handled in fibre glass, plastic, or stainless steel containers because it is *corrosive to the conventional metallic equipment.*

Copper complexes alone can also be used to control hydrilla but then it is to be applied in much higher concentration which can be damaging to the fish and other aquatic fauna. The mixture will also suppress scum algae and encourage phytoplanktons (Gupta, 1976).

In the running irrigation waters, **acrolein, xylene,** and **endothall** pellets have been found very effective in controlling hydrilla (Mehta, 1975). Of recent, a product called Clean-Flo-Lake cleanser has been reported capable of greatly weakening the hydrilla plants by precipitating phosphorus in its surrounding water.

Neither of the above mentioned herbicide treatments is capable of destroying tubers and turions of hydrilla burried in the hydrosoil. New hydrilla plants often come back within 30-45 days of their destruction with herbicides. But **fenac** (with or without amitrol), **dichlobenil, bensulfuron,** and **fluridone** are known to damage the hydrilla tubers. Single application of fenac at 34 kg/ha or of dichlobenil at 11 kg/ha kept hydrilla in check for 12 and 48 months, respectively (Steward, 1980). Controlling the hydrilla tubers is a major goal of aquatic plant managers which could seriously curtail the ability of hydrilla to re-infest the waterbody.

Biological control of hydrilla with **Chinese grasscarp** is the best choice for clearing water-ways of this menace. Hydrilla is on the food preference list of this bioagent. In some developing countries, including India, pouching of the animal by persons inhabiting around the treated waterbodies is a serious hurdle in employing grasscarp as a bioagent (Chapter 8).

In Pakistan a leaf mining fly *Hydrellia pakistanae* and a tuber boring weevil *Bagous affinis,* have been found efficient arthropod biocontrol agents of hydrilla. The latter operates only in very shallow waters or when the waterbody has dried due to dense growth of the weed. The earlier reported bioagent of hydrilla, *Parapoynx* sp., (a moth), was found consumed by fish present in the waterbodies and, therefore, it is no more considered useful in the control of hydrilla.

Amongst the possible fungal bioagents, *Fusarium roseum* strain *culmorum* has held promise in seriously infecting hydrilla.

3. *Ipomoea carnea* Jacq. **(Fig. 3.21)**

Biology and Habitat

I. carnea is a semi-woody plant which shows its initial presence as a weed in the neglected roadside and farm ditches where it forms, with the help of its ascending stems, a thick wall. In fact, because of this habit of the plant farmers used it as a biological fence plant to protect their farm boundaries. It is both drought and waterlogging resistant and is not consumed by farm animals As a result of all this, *I carnea* has become a very serious weed in many states of India where it chokes drainage and irrigation ditches, impedes water flow in canals by forming thickets in the shore waters, and creeps in many places into the paddy fields and other cultivable wet lands. Huge losses in agriculture are incurred in the country by this introduced plant.

Control Options

Practically no research work on the control of *I. carnea* has been reported in India so far. Some records from abroad show its high susceptibility to glyphosate and 2,4-D. Also, work on its biological control is in progress at Jabalpur and some other places. This weed, however, deserves much more attention at the hands of the weed scientists for its control than what it receives today.

4. *Nymphaea and Nelumbo* spp. (water-lily and lotus) **(Figs. 3.11-3.13)**

Biology and Habitat

Among the rooted floating aquatic weeds *Nymphaea stellata* Willd. and *Nelumbo nucifera* Gaerin are most troublesome species in India. They impound large areas of confined waters meant for pisciculture, drinking, irrigation, or recreation in Orissa,

Maharashtra, West Bengal, and Uttar Pradesh. In the famous Kashmir lakes, water-lily and lotus hinder boating.

Control Options

2,4-D, TCA, CSP, endothall, and dichlobenil have been found effective **herbicides** against water-lily and lotus. In India, *N. stellata* and *N. nouchali* were controlled in fish ponds with single application of **2,4-D** at 1.5 kg/ha mixed with 0.25-1% of a suitable wetting agent. The weedy plants were soon detached from their rhizomes after treatment whence these could be dragged to the shore with a rope. Ester 2,4-D proved more effective than its Na salt in this respect. Singh (1962) found lotus (*N. nucifera*) more susceptible to **TCA** than 2,4-D. In USA, **silvex** (LVE) was used against both water-lily and lotus. It was applied as 0.4 to 2.0 per cent foliage spray or at 20-40 kg/ha as through-the-water treatment. *Nelumbo* was also found susceptible to **endothall** at 0.5-1.5 ppm, **diquat** at 1 ppm, and **dichlobenil** at 6-8 kg ha^{-1}. Water-lily was controlled with foliar spray of **glyphosate** at 1-2 kg/ha.

Exposed bottom treatment of canals with dichlobenil at 5 kg/ha is also found highly effective against both, water-lily and lotus. Comes and Marrow (1971) reported that regrowth of water-lily from rhizomes was prevented with dichlobenil.

5. *Pistia stratiotes* L. (waterlettuce) **(Fig. 3.8)**

Biology and Habitat

P. stratiotes is a free-floating aquatic plant of tropical and subtropical waters. (Plate.6).Like waterhyacinth, it forms dense mats on the surface of water to hamper normal waterflow, navigation, and other water uses. Besides, it creates a favourable habitat for the growth of disease carrying mosquitoes (Pieterse, 1981).

Control Options

Waterlettuce mats are only 3% as heavy as the waterhyacinth mats. This makes their **drag-netting** comparatively easy.

Mineral oils, 2,4-D, paraquat, and anhydrous ammonia have been found effective **herbicides** against waterlettuce (Patnaik, 1972; Ramchandran *et al*., 1972). A **mineral oil** like kerosene is required in large quantities of about 1100 l/ha for effective kill of the weed. This is thus a very costly treatment. **2,4-D** (or MCPA) offers cheaper herbicidal control of waterlettuce. It is applied as a 0.2% spray, containing a suitable wetting agent. To avoid any spray drift hazard, invert emulsions of LVE of 2,4-D are preferred over its aqueous sprays.

Anhydrous ammonia is another very potent herbicide against waterlettuce. It can be either injected into the infested waterbody to achieve a concentration of 40 ppm N, or it may be applied as 1% foliar spray (with 0.25% of a wetting agent). About 70% of the sprayed waterlettuce plants succumb to NH_3 in 2 to 3 days.

Diquat (at 1-3 kg/ha) and **amitrol-T** (at 2-4 kg/ha) have also been found effective against waterlettuce (Anon, 1969).

6. *Potamogeton* spp. (pondweeds) (Figs. 3.25–3.28)

Biology and Habitat

Pondweeds infest both slow and fast moving waterbodies alike, obstructing their normal water flow, fishing, and aquatic sports. In India, the two most common pondweeds are: *Potamogeton pectinatus* (sago pondweed) and *P. nodosus* (long-leaf pondweed). These are reported weedy largely from Northern and North-Western states of India, as well as from Andhra Pradesh. In the country's famous Bhakra-Nangal and Chambal canal systems the pondweeds are the biggest nuisance.

Control Options

Solar treatment of exposed bottoms of canals and their **manual** clearance were found to keep the two species of pondweeds in check very temporarily.

Pondweeds have shown acceptable susceptibility to **herbicides** *e.g.* 2,4-D, chlorthiamid, acrolein, xylene, endothall, dichlobenil, and diuron. Dutta *et al.* (1972) found granular BEE of **2,4-D** (60 kg *a.e.*/ha) and **chlorthiamid** (16.5 kg *a.i.*/ha) very effective against these weeds. In irrigation waters, Mehta (1975) recommended the use of **acrolein** and **xylene** to control pondweeds. In the fishery waters, Hiltribran (1963) controlled pondweeds by underwater application of **endothall** (amine salt) at 1 ppmw. Repeat treatments were needed to check their regrowth. *P. nodosus* was controlled with 8 kg/ha **dichlobenil** applied through water. Pondweeds were also found susceptible to 0.5-1.5 mmpw **diquat** (Anon., 1969). In Illinois, **diuron** at 1 kg/ha eliminated *P. pectinatus* (and filamentous algae). To obtain continued suppression of the perennial pondweeds, exposed bottom treatment with dichlobenil at 4 kg/ha or fenac at 10-20 kg/ha was found useful.

Grasscarp is now a proven biocontrol agent of pondweeds. The adult carp consume all common species of pondweeds found in India although at its young age *P. perfoliatus* was preferred.

Adult grasscarp consumed daily 0.8-1.17 times its body weight in pondweeds (Micrewiez, *et al.* 1972). These authors noted that 20 grasscarp, each weighing about 348 g, reduced a pondweed infestation of 2 kg/sq m to nil in 34 days in a minor of Chambal canal.

7. *Salvinia molesta* Mitchell (waterfern, salvinia) **(Fig 3.4)**

Biology and Habitat

Of the seven or eight species of *Salvinia* on record in the tropics and subtropics, in India *S. molesta* and *S. natans* Allioni occur as weeds in her 20 districts. While *S. molesta* is common in the states of Kerala and Orissa, *S. natans* is found weedy in Jammu and Kashmir.

Salvinia originated in southeastern Brazil from where it spread widely, creating problems in Africa, Central and Southeast Asia, and the Southwest Pacific. In India, salvinia was introduced as a plant for botanical study. Today it is one of the worst aquatic weeds with us.

S. molesta is a free-floating aquatic fern but it can be found also in wet muds(Plate-11.7). Its mature infestation is called the *mat form* which is produced when the plants are crowded. The plant of this form of salvinia are relatively robust, with short internodes and overlapping pairs of conduplicately-folded, floating leaves, measuring up to 60 mm across when opened out. In comparatively less crowded conditions, the fern develops longer internodes and bears boat-shaped leaves. The leaf size then is 20-50 mm. This is called the *secondary form* of the plant. The *primary form* of salvinia occurs in new infestations of waterbodies. It bears flat, oval, up to 10-15 mm wide leaves. This form is also called the flat phase of the plant. The doubling time of salvinia is less than 3 days in mid summers and 40-60 days during winters. The plant is capable of luxury uptake of N and P from its surrounding water.

Salvinia reproduces and disperses vegetatively by its nodal buds and plant separation. Man, wind, and water currents are its chief dispersal agents.

In Kerala, *S. molesta* covered over 7000 sq km of water area, affecting daily life of about 5 million people. In several paddy fields it badly pulled down their production potentials.

Like many other weeds, salvinia mats reduce or prevent the use of water for recreation, fishing, and transport; interfere with floodgates, locks, and weirs; block drains; and degrades potable waters. In Orissa, *S. molesta* chokes rivers, lakes, and irrigation canals.

Plate 11.6: *Pistia stratiotes* L.

Plate 11.7: A dense infestation of *Salivinia molesta* Mitchell in Kerala.

Plate 11.6

Plate 11.7

Plate 11.8 : A thick stand of *Typha latifolia* L. badly blocking a drainage ditch in Chambal Command Area around Kota. (Photo-author)

Control Options

In small areas, physical **netting** is found effective in India. The operation is best conducted during March to May when the fern mats are weak. Also, physical **flushing** of salvinia from lagoons to the main stream, from where the weed is eventually washed out to sea, is sometimes helpful in providing at least a short time relief.

The manually lifted salvinia plants have sometimes been proposed for use as mulch in plantations and for composting for use in agricultural fields. But such benefits of salvinia should be weighed against its dissemination potentials. Perhaps, a better alternative is to **utilise** salvinia as raw material for making paper, carboard, fibreboard, and road stabilisers since the fern is rich both in its fibre and ash contents. A factory to this effect is proposed in Kerala. In no case salvinia should be used as packing material which has been an important source of its dispersal in India.

Several **herbicides** have been tried against *S. molesta.* but only a few killed it to acceptable levels. The hydrophobic surfaces of salvinia leaves do not permit their wetting with the herbicide sprays. Also, even when the weed is found susceptible to a herbicide, the application of the herbicide in salvinia-infested waterbody is not easy. Because of such reasons, herbicidal control of salvinia has never gone to the field.

Initially, a mixture of **kerosene,** diuron, and surfactant (AF-101) was found to destroy the leaf papillae of salvinia and, thus, sank the salvinia mats. **PCP** (alongwith a wetting agent) was found successful in Ceylon in destroying salvinia in lowland paddy fields when applied about a week before transplanting rice seedlings. **Paraquat** can also be tried in this situation against the fern (Thomas, 1985).*In other waterbodies, PCP has limited utility because of its toxicity to fish, frogs, and tortoise.*

In paddy fields, **thiram** (a fungicide) and **niclosamide** (a molluscocide) have been reported to cause selective control of salvinia. In Argentina, complete control of this weed was achieved with **hexazinone** at 1.1-2.25 kg/ha. The control lasted for about one year. **Ametryn** (0.4-1.6 ppm) has also been found effective against salvinia.

In the last decade or so, the **biological control** of *S. molesta* with a small weevil, *Cyrtobagous salviniae* Sands (Curculionidae) of South American origin has come to the rescue of people of Kerala. The weevil was first reported as a potential bioagent against *S. molesta* in Australia in 1981 and later in Kerala (India) in 1984 (Room *et al.,* 1981 and Joy *et al.* 1984).

In India, in 1982 about 4000 *Cyrtobagous* weevils were introduced from Australia to control *S. molesta* at different sites in Kerala. The weevil established itself in all the released sites and a number of infested ponds and canals were cleared of this weed within

14 months. Also, it spread to nearby rice fields controlling the weed there without damage to rice plants. Likewise, in Banglore a water-lily pond showing 75% infestation of salvinia was completely freed of this fern with *Cyrtobagous* weevils.

A characteristic feature of *C. salviniae* is its very lazy dispersal nature. It is so because it rarely flies. But the weevil is strong at its establishment; one female in a lot being able to establish a population. Because of its very slow dispersability, it is recommended to distribute the weevil-infested salvinia plants in small lots over the entire waterbody under treatment. Tidal waves and navigation are helpful in augmenting dissemination of the weevil. The weevil takes about 4-6 months to turn the infected salvinia mats yellow and another 6 months to sink them.

It has been reported that pre-treatment of salvinia with kerosene (alongwith a surfactant) can help the *C. salviniae* weevil to feed upon the weed fast (Julien and Storrs, 1996).

8. *Typha* spp. (cattails) **(Figs. 3.58 - 3.59)**

Biology and Habitat

Cattails form the major ditch and shoreline weeds in India, particularly in the states of Rajasthan, Haryana, Uttar Pradesh, Kashmir, Madhya Pradesh, Karnataka, and Andhra Pradesh. These robust, grasslike plants form dense clumps by rapid growth of their rhizomes and invade new ditches and shores by their air borne seeds (Plate 11.8). Each cattail shoot may produce up to 20,000 seeds. They germinate readily on wet muds.

The cattail seedlings cannot withstand continuous submergence, but once established these can put forth heavy growth in flooded situations. Cattails from seeds produce spikes and rhizomes during the second and subsequent years of their growth. In a single season an old *Typha* shoot can spread its rhizomes 3 m around in all directions, producing 100 or more clumps in the process.

Cattails grow both in fresh and brackish waters. In India, *Typha angustata* (narrow leaf cattail) is the major cattail species, but *T. latifolia* (broadleaf cattail) and *T. elephantina* also act as weeds in many places. While *T. angustifolia* grows in deep waters, *T. latifolia* is a shallow water plant.

Control Options

Mowing of *Typha* with swords and sickles is practised in India but the regrowth from the cut plants is very rapid. It is found that if the flowering shoots of *Typha* are cut below the water level, its underwater biomass begins to decay rapidly. The decay process raises the B.O.D. of the water system so much that it induces anaerobic respiration in the cattail roots, which further hastens their self destruction and postpones the cattail regrowth for 6 months or so (Sale & Wetzel, 1983). Usually, three underwater mowings of cattails during the monsoon season suffice their destruction.

2,4-D was the first **herbicide** found successful against *Typha* spp. A drenching spray of **2,4-D** (LVE) with diesel oil (at 4-8 kg *a.e.*/ha) at the preflowering stage of cattails has long been used in the past. Usually, repeat applications for 3 consecutive years are required to recover the area from *Typha* spp. **Dalapon** and **amitrole,** although costlier than 2,4-D, have been found more effective against cattails. These two herbicides are not required to be sprayed in drenching volumes and, at the same time, are much less prone to drift hazards than 2,4-D. The common rates of dalapon and amitrole are, respectively, 15-30 kg/ha and 6-12 kg/ha. But it is often better to consider these application rates in terms of spray concentrations since in ditches the area is difficult to measure. The effective concentrations are 2.5% for dalapon and 1.25% for amitrole. In either case, one may add a suitable surfactant to improve the weed kill. Sometimes, these herbicides are applied as emulsion sprays with 5-10% of diesel oil. Fully-headed and post-heading stages of cattails are found most susceptible to dalapon and amitrole (Anon., 1969).

A **tank mixture** of 15 kg dalapon and 3 kg amitrole per ha has been found more effective than either herbicide applied alone against the cattails. The mixture injured even some shallow rhizomes of the weed. Usually, two sprays are needed to obtain their complete kill. The first spray is applied when the cattails are in full spike and the second one as a spot treatment to deal with the escapes. Mehta and Boonlia (1984) further reported that dalapon (25 kg ha^{-1}) and amitrole (8 kg ha^{-1}) worked better against cattails when these were sprayed on three months' old regrowth of the mowed weed. For this treatment the optimum periods for herbicide application were July in case of dalapon and September to January for amitrole.

Glyphosate at 2-3 kg/ha is another very good translocated herbicide against *Typha* spp. The herbicide works best at the full bloom stage of the weed. Delayed spray at its fruiting stage would allow some cattails to regenerate from seeds.

Diquat and **paraquat** are also found very effective on cattails. But these, being contact herbicides, are preferred when a rapid kill of the weed shoots is needed. Their effective doses are 1.2-1.6 kg/ha (or 0.22% spray). The treatment will leave the lower most portions of cattails unhurt inside ditches.

Mehta and Sharma (1975) demonstrated that on wet lands recovered from *Typha* spp., *Brachiaria mutica* (paragrass) slips could be planted to prevent their reinfestation with this weed. Paragrass was found to establish quickly to provide green and nutritious fodder within four months.

9. *Vallisneria spiralis* L. (eelgrass) **(Fig. 3.38)**

Biology and Habitat

Eelgrass is a submersed, rooted weed of the confined or slow moving waterbodies. Its chief obnoxious features are: hindrance in healthy aquatic sport and reduction in water flow rates. In India it is reported problematic chiefly from Rajasthan, Haryana, Uttar Pradesh, and Andhra Pradesh.

Control Options

Amine salts and esters of **2,4-D** are found effective herbicides against *Vallisneria*. Their effective rates are 6 kg/ha or 20-60 ppmw concentration in the waterbody (Dutta and Gupta, 1976). The weed is also found susceptible to **CSP** at 2 ppm concentration. It killed the weed in 25 days. A **mixture** of CSP and fenac (at 1 ppm + 2.5 ppm) greatly expedited the control of eelgrass (Mehta and Boonlia, 1975).Application of mud pellets containing CSP (90 kg ha^{-1}) in three splits prevented the new growth of eelgrass in still waterbodies (Mitra and Banerjee, 1968). The weed has also been found suceptible to endothall amine at 6 ppm (Mehta, 1975).

Grasscarp is a rapid feeder of eelgrass. Both young and adult individuals of grasscarp use eelgrass as their priority food. Its average daily consumption by the carp is 0.94-1.19 times its body-weight.

SELECTED BIBLIOGRAPHY

1. Anon. 1969. Control of aquatic weeds. USDA Agr. Handbook No. 33, 60-64 pp.
2. Bhatia, H.L. 1970. Grasscarp can control aquatic weeds. Ind. Fmg. 20 (2): 36-37.
3. Bowes, G. *et al*. 1977. Adaption to low light levels by hydrilla. J. Aquat. Plant Manage. 15 (June): 32-35.
4. Cary, P.R. and P.G.J. Weerts. 1983. Growth of *Salvinia molesta* as affected by water temperature and nutrition. Aquat. Bot. 16 (2): 163-172 and 17: 61-70.
5. Center, T.D. 1982. The waterhyacinth weevils. Aquatics 4 (2): 8.
6. Center, T.D. 1984. Dispersal and variation in infestation intensities of waterhyacinth moth *Sameodes albiguttalis* (Lepidoptera: Pyralidae) populations in Penninsular Florida. Enviro. Ento. 13 (2): 482-491.
7. Center, T.D. and N.R. Spencer. 1981. The phenology and growth of waterhyacinth (*Eichhornia crassipes* Mart. Solms) in a Eutrophic North-Central Florida lake. Aquat Bot. 10: 1-32.

8. Center, T.D. *et al.* 1982. Control of waterhyacinth with *Neochetina eichhorniae* and a growth retardant. Weed Sci. 30 (5): 453-457.

9. Center. T.D. and W.C. Durden. 1981. Release and establishment of *Sameodes albiguttalis* for the biological control of waterhyacinth. Environment Ento. 10(1): 75-80.

10. Chibbar, S.S. and G.D. Singh. 1971. Paddy straw and waterhyacinth silage. Ind. Fmg. 20 (11): 24-26.

11. Comes, R.D. and L.A. Marrow. 1971. Control of water-lilies with dichlobenil. Weed Sci. 19: 402-405.

12. Cook, C.D.K. and B.J. Gut. 1971. Salvinia in the state of Kerala (India). PANS 17 (4): 438-447.

13. Corns, W.G. and R.K. Gupta. 1971. Chemical control of the cattail (*Typha latifolia).* Canadian J. Plant Sci. 51 (6): 491-497.

14. Dutta, T.R. and J.N. Gupta 1976. Some experiments on aquatic weed control in fisheries lakes and streams in U.P. (*IN*) Aquatic weeds in S.E. Asia, Dr. W. Junk b.v. Pub., The Hague, 249-252 pp.

15. Dutta, T.R. *et al.* 1972. Evaluation of herbicides for submersed weeds in Chambal and Bhakra Nangal Canal Systems. Ind. J. Agr. Sci. 42 (1): 70-75.

16. Eberbach, P.L. and K.H. Bowmer. 1995. Conversion of ^{14}C-glyphosate to carbon dioxide by alligator weed. J. Aquatic Plant Manage. 33, 27-29.

17. Finlayson, C.M. and T.P. Farrell. 1983. Chemical control of *Salvinia molesta*. Aust. Weeds 2 (3): 90-93.

18. Gangstad, E.O. *et al.* 1975. Towards integrated control of alligator weed. Hyacinth Control J. 13 (June): 30-38.

19. Gibson, M.T. *et al.* 1990. Barley straw as inhibitor of algal growth. J. Applied Physiology 2 (3): 241-248.

20. Gopalkrishnan, R. and P.J. Joy. 1977. Evidence that N supply influences the distribution of a freshwater macrophyte *Ceretophyllum demersum*. J. Ecol. 59, 783-791.

21. Gupta, O.P. 1976. Chemical control of hydrilla (*Hydrilla verticiallata* Casp.) (Hydrocharitaceae). Rajasthan J. Pesticides 3 (1 & 2): 18-24.

22. Gupta, O.P. 1978. Effect of high sodium adsorption ratio of water on the growth of waterhyacinth (*Eichhornia crassipes*). Inst. Agr. and Animal Sci. J. 1 (2): 54-59.

23. Guritno, B. 1975. The growth of *Salvinia molesta* D.S. Mitchell with special reference to the effect of nutrient factors. Biotrop. News Letter (Indonesia) No. 11, 12.

24. Haller W.T. and D.L. Sutton. 1973. Effect of pH and high P concentrations on growth of waterhyacinth. Hyacinth Contr. J. 11 (June) 59-61.

25. Hart, M.N. and K.M. Srinivasan, 1966. Observations on the comparative effectiveness of aminotriazole and 2,4,5-T amine in the control of waterhyacinth (*Eichhornia crassipes*). Mad. Agri. J. 53: 289-292.

26. Hiltibran, R.C. 1963. Effect of endothall on aquatic plants. Weeds 11 (4): 256-257.

27. Hodgson, L.M. and C.C. Certer. 1982. Effect of hydrilla management by herbicides on a periphyton community. J. Aquat. Plant Management 20: 17-19.

28. Jagadeesh, K.M. and C.S. Lakshminarayana. 1971. Eradication and utilisation of waterhyacinth-A review. Curr. Sci. 40 (7): 148-149.

29. Jain, S.C. 1975. Aquatic weeds and their management in India. Hyacinth Control J. 13 (June): 6-7.

30. Jordon, L.S. *et al.* 1962. Chemical control of filamentous green algae. Hilgardia 32 (9): 433-441.

31. Joy, P.J. 1978. Ecology and control of *Salvinia* (African Payal), the molesting weed of Kerala. Tech. Bull. No. 2., Kerala Agri. Univ., 49 p.

32. Joy, P.J. *et al.* 1984. Establishment of *Cyrtobagous salviniae* Sands on *Salvinia molesta* Mitchell in Kerala. Agri. Res. J. Kerala 22 (1): 105-106.

33. Joyce, J.C. and W.T. Haller. 1984. Effect of 2,4-D and giberellic acid on waterhyacinth under operational conditions J. Aquat. Plant Manage. 22: 75-78.

34. Julien, M.H. and Storrs, M.J. 1996. Integrating biological and herbicidal controls to manage *Salvinia* in Australia. Weed Abst. 45 (12): 4996.

35. Kammathy, R.V. 1968. *Salvinia auriculata* Aublet-A rapidly spreading exotic weed in Kerala. Sci. Cult. 34 (9): 396.

36. Kathiresan, R.M. 1999. Use of bioagents in weed management. Souvenir 8th Biennial conference of the Ind. Soc. Weed Sci. held 5th-7th Feb.99 at Varanasi (India): 41-46.

37. Kaul, V. and U. Bakaya. 1976. The noxious floating *Lemnids-Salvinia* aquatic weed complex in Kashmir. (*IN*) Aquatic Weeds in S.E. Asia, Dr. W. Junk b.v. Pub. The Hague, 183-192 pp.

38. Langeland, K.A. 1995. *Hydrilla verticillata*- the prefect aquatic weed. Castanea 61 (3): 293-304.

39. Leitch, J.A. *et al.* 1997. Economics of cattail control to reduce blackbird damage to sunflower. Agriculture, Ecosystems, and Environ. 65: 141-149.

40. Lindsey, K. and H.M. Hirt. 1999. Use Waterhyacinth ! A Practical handbook of Uses for the Waterhyacinth from Across the World. Anamed, Schafweide 77, 71364 Winnenden, Germany, 115 p.

41. Loyal, D.S. and R.K. Grewal. 1966. Cytological study on sterility in *Salvinia auriculata* Aublet. Cytologia 31: 330-338.

42. McGilverey, F.B. and J.H. Steenis. 1965. Control of Alligatorweed in South Carolina with granular silvex. Weeds 13, 66-68.

43. Mehta, I. 1975. Some experiments with xylene, acrolein, and Hydrothol-191 for the control of submersed weeds. Andhra Agri. J. 22 (1 & 2): 20-24.

44. Mehta, I. and D.S. Boonlia. 1975. A note on chemical control of *Hydrilla and Vallisneria* spp. Ind. J. Weed Sci. 7 (2): 125-126

45. Mehta, I. and D.S. Boonlia. 1980. Weed out weed with a weed. Intensive Agri. 18 (6): 23-24.

46. Mehta, I. and D.S. Boonlia. 1984. Effect of seasonal sprays of dalapon and aminotriazole on the population fluctuations of *Typha angustata*. Ind. J. Weed Sci. 16 (1): 6-10.

47. Mehta, I. and R.K. Sharma. 1975. Control of *Typha* by competitive plants. Annals Arid Zone 14 (2): 175-182.

48. Micrewiez, J.E. *et al.* 1972. The whiteamur for aquatic weed control. Weed Sci. 20 (1): 106-110.

49. Misra, A. *et al.* 1976. Studies on chemical control of Chara. (*IN*) Aquatic Weeds in S.E. Asia, W. Junk b.v. Pub., The Hague 265-268.

50. Mitra, E. and A.C. Banerjee. 1968. Release of copper ions from pond soils. Ind. Agric. 12: 192-193.

51. Mukherji, S.K. 1970. Further studies on the chemical control of algal weeds. World Crops 22: 387-388.

52. Mukherji, S.K. 1972. Use of pentaphenol as an algaecide in paddy fields in W. Bengal. Weed Res. 12 (4): 389-388.

53. Mukherji, S.K. and S.K. Sengupta. 1964. Control of algal weeds in paddy fields in West Bengal (India). FAO Plant Protection Bull. 12: 129-130.

54. Muniyappa, T.V. 1983. Preliminary studies on the use of paraquat and 2,4-D for the control of waterhyacinth. Ind. J. Weed Sci. 15 (2): 243-246.

55. Olaleye, V.F. and O.A.A. Yemuu. 1996. Effect of glyphosate application to control *Eichhornia crassipes* on fish composition. J. Environ. Manage. 47 (2): 115-122.

56. Ori, Y. 1994. Integrated management of aquatic weeds in Japan. Weed Abst. 45 (4), 1643.

57. Patnaik, S. 1972. Observations on the control of *Pistia stratiotes* in fish ponds by gramoxone. Proc. Ind. Acad. Sci. (B) 75 (1): 15-22.

58. Philipose, M.T. 1963. Making fish ponds weedfree. Ind. Livestock 2: 20-21 & 34-36.

59. Pieterse, A.H. 1981. *Hydrilla verticillata*-A review. Abst. Trop. Agri. 7 (6): 9-34.

60. Pieterse, A.H. and F.A. Roorda. 1982. PGR sprays boost 2,4-D activity against waterhyacinth. Aquat. Bot. 13 (1): 69-72.

61. Ramachandran, V. 1960. Observations on the use of ammonia for the eradication of aquatic vegetation. J. Sci. & Indust. Res. 19 (I): 284-285.

62. Ramachandran, V. 1969. Weed control is a must in fish culture. Ind. Fmg. 19 (12): 4.

63. Ramachandran, V. *et al.* 1972. Observations on the use of ammonia for the control of *Pistia stratiotes* Linn. J. Inland Fish Soc.: 124-130.

64. Riemer, D.N. 1976. Long term effects of glyphosate applications to *Phragmites.* J. Aquat. Plant Manage. 14 (6): 39-42.

65. Room *et al.* 1981. Biocontrol of *Salvinia* in Australia. Nature 294.

66. Room, P.M. *et al.* 1981. Successful biological control of the floating weed, salvinia. Nature 294: 78-80.

67. Sale, P.J.M. and R.G. Wetzel. 1983. Growth and metabolism of *Typha* in relation to cutting treatments. Aquat. Bot. 15 (3): 321-334.

68. Singh, D.J.C. and K.N. Rao. 1976. Chemical control of the aquatic weed *Typha.* Pesticides 10 (8): 45.

69. Singh, H.P. and S.N. Sharma. 1993. Efficacy of dalapon and 2,4-D on chlorophyll content of two emergent weeds. Agri. Sci. Digest (Karnal), 13 (2): 99-103.

70. Singh, S. and M.S. Gill. 1996. Evapotranspiration loss of waterhyacinth and chemical control of weed. Ind. J. Agri. Sci. 66 (8): 487-489.

71. Steward, K.K. 1969. Effect of growth regulators and herbicides on the germination of hydrilla turions. Weed Sci. 17 (3): 299-301.

72. Steward, K.K. 1980. Retardation of *Hydrilla verticillata* regrowth through chemical control of vegetative reproduction. Weed Sci. 28: 245-251.

73. Steward, K.K. 1993. Seed production in monoecious and dioecious populations of *Hydrilla.* Aquatic Botany 46 (2): 169-183.

74. Sutton, D.K. *et al.* 1970. Effect of diquat on uptake of copper in aquatic plants. Weed Sci. 18 (6): 703-770.

75. Swarbrick. J.T. 1982. The biology and control of *Hydrilla verticillata*. Biotrop Special Pub. (Indonesia), 16.

76. Thomas, P.A. 1985. Management of *Salvinia molesta* in Papua New Gunea. FAO Plant Protection Bull. 33 (2): 52-56.

77. Timmons, F.L. *et al.* 1958. A study of factors which influence effectiveness of amitrole and dalapon on common cattail (*T. latifolia*). Weeds 6: 406-412.

78. Van, T.K. and T.D. Center. 1994. Effect of pactobutrazol and waterhyacinth weevil (*Neochetina eichhorniae*) on plant growth and leaf dynamics of waterhyacinth. Weed Sci. 42 (4): 665-672.

79. Van, T.K. and V.V. Vandiver. 1992. Response of monoecious and dioecious hydrilla to bensulfuron methyl. J. Aq. Plant Manage. 30, 41-44.

80. Visalakshy, P.N.G. and K.P. Jayanth. 1996. Effect of silt coverage of waterhyacinth roots on pupation of *Neochetina eichhorniae* and *N. bruchi*. Biocontrol Sci. and Technol. 6 (1): 11-13.

81. Westbrook, R.C. 1984. Fedral noxious weeds--Kariba weed. Weeds Today 15 (1): 8-9.

82. Yeo, R.R. and N. Dechoretz. 1977. Acute Toxocity of a herbicidal combination of diquat plus copper ion to eggs. J. Aquat. Plant Manage. 15 (6): 57-59.

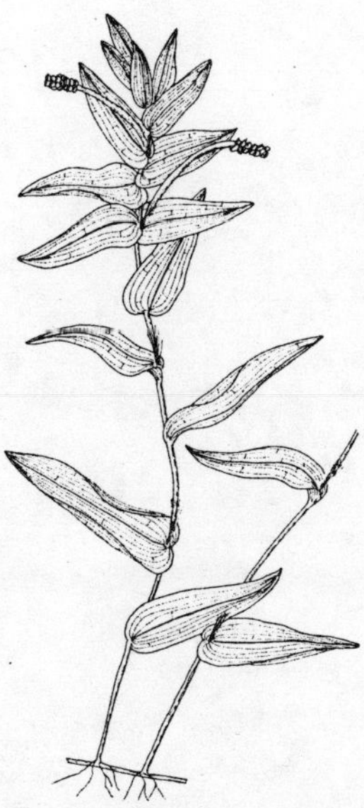

Potamogeton richardsonii (Richadson pondweed)

Appendix 1

AQUATIC PLANT JOURNAL CITATIONS

Information about aquatic plants and their management is found in many journals. Most of the research articles are found in about 150 journals published from different countries, although sporadic references are available also in another 200 journals or so. Of these journals, below are listed some major journals in the two groups, in *alphabetic order*.

GROUP-I

1. AMERICAN JOURNAL OF BOTANY
   Botanical Society of America Inc.
   Ohio State University Department of Botany
   1735 Neil Avenue,
   Columbus, Ohio 43210 (USA)

2. AQUATIC BOTANY
   Elsevier Scientific Publication Co.,
   P.O. Box 330
   1000 AH Amsterdam
   (Netherlands)

3. AQUAPHYTE (NEWS BULLETIN)
   Centre for Aquatic and Invasive Plants,
   7922 N.W. 71st Street
   Gainsvilla, Florida-32653-3071 (USA)

4. CANADIAN JOURNAL OF BOTANY
   National Research Council
   Ottawa, Ontario KIA OR 6
   (Canada)

5. HYDROBIOLOGIA
   Dr. W. Junk Publishers
   P.O. Box 13713
   2501 E.S. The Hague
   (Netherlands)

6. JOURNAL OF AQUATIC PLANT MANAGEMENT
   Aquatic Plant Management Society
   P.O. Box 16
   Vicksburg, Mississippi 39180 (USA)

7. JOURNAL OF ECOLOGY
   Blackwell Scientific Publications Ltd.
   Osney Mead
   Oxford OX 2 OEI
   (United Kingdom)

8. WEED SCIENCE
   Weed Science Society of America
   309 West Clark Street
   Champaign, Illinois 61820 (USA)

GROUP-II

1. Fisheries Management
2. Freshwater Biology
3. Hyacinth control J.
4. Hydrological Bull.
5. Indian J. Marine Sci.
6. Indian J. Weed Sci.
7. J. Freshwater Eco.
8. Japanese J. Limnology
9. J. Fishery Research Board, Canada
10. J. Water Pollution Control Fed.
11. Limnol. Oceanograph
12. Marine Biol.

13. Progressive Fish-culturist
14. Royal J. Aquatic Plant Management
15. Tropical Pest Management
16. Water Research
17. Water Resource Bull.
18. Water Research
19. Water Supply
20. Watsonia
21. Weeds
22. Weed Research

Appendix 2

SOME CONTACT SOURCES FOR FURTHER INFORMATION ON AQUATIC WEEDS

1. Agricultural Experiment Station, University of New *Hampshire*, Durhan, Hampshire (England)
2. Aquatic Weeds Research Center, University of Florida, Ft. Lauderdale, *Florida*, (USA)
3. Bureau Aquatic Plant Research and Control, Florida Dept. Nat. Res., *Florida*, (USA)
4. Central Inland Fisheries Res. Instt. *Barrackpore*, West Bengal (India)
5. Council of Scientific and Industrial Research Organisation, Division of Irrigation Research, Griffith, *New South Wales* 2680 (Australia.)
6. Department of Entomology, Assam Agricultural University, *Jorhat*-785013 (India)
7. International Plant Protection Centre, Center for Aquatic Weeds, University of Florida, *Gainesville* (USA)
8. Project Directorate of Biological Control, Bellary Road, Hebbal Agricultural Farm Post, *Bangalore* – 560024 (India)
9. U.S.D.A., A.R.S., Aquatic Plant Management Lab., Fort. Lanuderdale, *Florida*, USA)
10. Weed Research Organisation of Agricultural and Forestry Research Council (C.A.B. Begbroke Hill, Yarton, *Oxford*, (England)

Appendix - 3

CHEMICAL NAMES OF AQUATIC HERBICIDES MENTIONED IN THE BOOK

ACROLEIN	=	2-propenal
AMETRYN	=	*N*-ethyl-*N'*-(1-methylethyl) - 6-(methylthio)-1,3,5-triazine-2,4-diamine
AMITROLE	=	1 *H*-1,2,4 - triazol-3-amine
AMITROLE-T	=	Amitrole+ammonium thiocyanate
BRESTAN	=	Triphenyl tin acetate
BUTHIDAZOLE	=	3-[5-(1,1-dimethylethyl)-1,3,4-thiadiazol-2-yl]-4-hydroxy-1-methyl-2-1 midazolidinone
CHLORSULFURON	=	2-chloro-*N*-[[(4-methoxy-6-methyl-1,3,5-triazin-2-yl)amino] carbonyl] benzenesulfonamide
2,4-D	=	(2,4-dichlorophenoxy)acetic acid
DALAPON	=	2,2-dichloropropionic acid
DASM	=	Dihydroxy aluminium (salt of) methylarsonic acid
DICHLOBENIL	=	2, 6-dichlorobenzonitrile
DSMA	=	Disodium (salt of) methylarsonic acid
DICHLONE	=	2, 3-dichloro-1,4-naphthoquinone

DIQUAT	=	6,7-dihydrodipyrido[-1,2- α : 2',1'-c] [pyrazinedium ion
DIURON	=	3(3,4 - dichlorophenyl)-1,1-dimethylurea
ENDOTHALL	=	7-oxyabicyclo [2.2.1] heptane-2, 3-dicarboxylic acid
FENAC	=	2,3, 6-trichlorbenzeneacetic acid
FENURON	=	*N*, *N*-dimethyl-N^1-phenylurea
GLYPHOSATE	=	*N* - (phosphonomethyl) glycine
HEXAZINONE	=	3-cyclohexyl-6-(dimethylamino)-1-methyl-1, 3 5- triazine-2,4(1*H*, 3*H*)-dione
MCPA	=	4(4-chloro-2-methylphenoxy) acetic acid
METHAM (SMDC)	=	Methylcarbamodithioic acid
MH	=	1,2-dihydro-3, 6-pyridazinedione
MONURON	=	N^1-(4-chlorophenyl)-*N*,*N*-dimethylurea
NABAM	=	di-sodium ethylene bio-dithiocarbamate
NITROFEN	=	2,4-dichlorophenyl-4-nitrophenyl ether
PARAQUAT	=	1,1'-dimethyl-1, 4'-biprodinium ion
PCP	=	Pentachloro phenol
SILVEX (=FENOPROP)	=	($\pm$)-2-(2, 4, 5-trichlorophenoxy) propionic acid
SIMAZINE	=	6-chloro-N^2, N^4-diethyl-1, 3, 5-triazine-2, 4-diamine
TERBUTRYN	=	N^2-*tert* butyl-N^4-ethyl-6-methylthio-1,3,5-triazine-2, 4-diamine.
TCA	=	Trichloroacetic acid

Appendix - 4

PRIVATE ORGANISATIONS DEALING WITH HERBICIDES IN INDIA

1. Aventis Crop Science
   Hoechst Centre
   54/A, Andheri Kurla Road,
   Andheri (E), Mumbai-400093

2. Bayer (India) Ltd.
   Bayer House, Central Avenue
   Hiranandani Garden
   Powai, Mumbai-400076

3. Cyanamid Agro Ltd,.
   4th Floor,
   16, Maharishi Karve Road
   Churchgate, Mumbai-400020

4. De-Nocil Crop Protection Ltd.
   I & II Floor, Administrative Building
   Pirojsha Nagar,
   Eastern Exports Highway,
   Vikhroli (E)
   Mumbai - 400079

5. E.I. Du Pont India Ltd.
   DLF Plaza Tower
   8th Floor, DLF Qutab Enclave, Phase-I
   Gurgaon-122002 (Haryana)

6. Gharda Chemicals Ltd.
   B-27/29, MIDC, Dombivil (E)-421203

7. Indian Crop Protection Association
 102, Creative Industrial Building,
 Sundernagar
 Road No.2, Kalina
 Santacruz (E)
 Mumbai-400098

8. Indofil Chemicals Company
 Nirlon House
 Dr. Annie Beasant Road
 Mumbai-400025

9. Monsanto Chemicals of India Ltd.
 Agriculture Sector,
 Ahura Centre,
 5th Floor
 96 Mahakali Caves Road
 Andheri (East)
 Mumbai-400093

10. Rallis Research Centre
 21 & 22 Peenya Industrial Area Phase-II
 Bangalore-560058

Addenda

ADDITIONAL REFERENCES

1. Cillers, C.J. 1999. Biological control of parrot's feather, *Myriophyllim aquaticum* (Vell.) Verdc. (Haloragaceae) in South Africa, 1999. African Entomol. Memoir 1 : 113-118.

2. Daou, H. and R.E. Talbert. 1999. Control of propanil-resistant barnyardgrass (*Echinochloa crus-galli*) in rice (*Oryza sativa*) with carbaryl/propanil mixtures. Weed Technol. 33 : 65-70.

3. Gopalaswamy, G. and S. Kannaiyan. 1998. Utilization of *Azolla* hybrids for bio-reclamation of sodic soil and rice (*Oryza sativa*) growth. Ind. J. Agron. 43(3) : 511-517.

4. Julien, M.H. *et al.* 1999. Biological Control of Waterhyacinth – The Weevils *Neochetina bruchii* and *N. eichhorniae.* ACIAR Monograph No. 60, Australia, 87 p.

5. Lee, C.L., *et al.* 1999. Heavy metals removal by a promising, locally available aquatic plant, *Najas graminea* Del. In Taiwan. Water. Sci. Technol. 39 (10-11) : 177-181.

6. Maceina, M. and J. Slipke. 1999. Confining grasscarp with an electric barrier within an embayment in Lake Seminole. Aquatics 22 (3) : 4, 7-9.

7. Prescott, G.W. 1951. Algae of the Western Lakes Area. Cranbrook Institute of Science Bull. No. 31, 946p.

8. Stocker, R.K. and W.T. Haller. 1999. Residual effects of herbicide – treated *Eichhornia crassipes* used as a soil amendment. Hydrobiologia 415 : 329-334.

9. Sutton, D.L. and V.V. Vandiver Jr. 1998. Grass Carp – A Fish for Biological Management of *Hydrilla* and Other Aquatic Weeds in Florida. APIRS, Florida (USA) Bull. No. 867, 13 p.

10. Zha, M. *et al.* 1999. Removal and recovery of zinc from solution and electroplating effluent using *Azolla filiculoides.* Water Research 33 (6) : 1516-1522.

Subject Index

R

S

Man making furniture from waterhyacinth under WHUP, USA.